T0137076

Intelligent Systems Reference Library

Volume 151

Series editors

Janusz Kacprzyk, Polish Academy of Sciences, Warsaw, Poland
e-mail: kacprzyk@ibspan.waw.pl

Lakhmi C. Jain, Faculty of Engineering and Information Technology, Centre for Artificial Intelligence, University of Technology, Sydney, NSW, Australia;
Faculty of Science, Technology and Mathematics, University of Canberra, Canberra, ACT, Australia;
KES International, Shoreham-by-Sea, UK
e-mail: jainlakhmi@gmail.com; jainlc2002@yahoo.co.uk

The aim of this series is to publish a Reference Library, including novel advances and developments in all aspects of Intelligent Systems in an easily accessible and well structured form. The series includes reference works, handbooks, compendia, textbooks, well-structured monographs, dictionaries, and encyclopedias. It contains well integrated knowledge and current information in the field of Intelligent Systems. The series covers the theory, applications, and design methods of Intelligent Systems. Virtually all disciplines such as engineering, computer science, avionics, business, e-commerce, environment, healthcare, physics and life science are included. The list of topics spans all the areas of modern intelligent systems such as: Ambient intelligence, Computational intelligence, Social intelligence, Computational neuroscience, Artificial life, Virtual society, Cognitive systems, DNA and immunity-based systems, e-Learning and teaching, Human-centred computing and Machine ethics, Intelligent control, Intelligent data analysis, Knowledge-based paradigms, Knowledge management, Intelligent agents, Intelligent decision making, Intelligent network security, Interactive entertainment, Learning paradigms, Recommender systems, Robotics and Mechatronics including human-machine teaming, Self-organizing and adaptive systems, Soft computing including Neural systems, Fuzzy systems, Evolutionary computing and the Fusion of these paradigms, Perception and Vision, Web intelligence and Multimedia.

More information about this series at http://www.springer.com/series/8578

Leslie F. Sikos
Editor

AI in Cybersecurity

Editor
Leslie F. Sikos
School of Information Technology
and Mathematical Sciences
University of South Australia
Adelaide, SA, Australia

ISSN 1868-4394 ISSN 1868-4408 (electronic)
Intelligent Systems Reference Library
ISBN 978-3-030-07539-2 ISBN 978-3-319-98842-9 (eBook)
https://doi.org/10.1007/978-3-319-98842-9

© Springer Nature Switzerland AG 2019
Softcover re-print of the Hardcover 1st edition 2019
This work is subject to copyright. All rights are reserved by the Publisher, whether the whole or part of the material is concerned, specifically the rights of translation, reprinting, reuse of illustrations, recitation, broadcasting, reproduction on microfilms or in any other physical way, and transmission or information storage and retrieval, electronic adaptation, computer software, or by similar or dissimilar methodology now known or hereafter developed.
The use of general descriptive names, registered names, trademarks, service marks, etc. in this publication does not imply, even in the absence of a specific statement, that such names are exempt from the relevant protective laws and regulations and therefore free for general use.
The publisher, the authors and the editors are safe to assume that the advice and information in this book are believed to be true and accurate at the date of publication. Neither the publisher nor the authors or the editors give a warranty, express or implied, with respect to the material contained herein or for any errors or omissions that may have been made. The publisher remains neutral with regard to jurisdictional claims in published maps and institutional affiliations.

This Springer imprint is published by the registered company Springer Nature Switzerland AG
The registered company address is: Gewerbestrasse 11, 6330 Cham, Switzerland

Foreword

Artificial intelligence (AI) is a relatively mature research topic, as evidenced by the top two AI conferences, the International Joint Conference on Artificial Intelligence and the AAAI Conference on Artificial Intelligence, which have been held since 1969 and 1980, respectively. Over the years, the advances in AI techniques have been phenomenal, and now AI is nearly ubiquitous. For example, the Guardian recently reported that AlphaZero AI beats champion chess program after teaching itself in 4 hours. Thus, a world in which everything is influenced, even to the extent of being controlled, by AI is no longer science fiction.

AI has applications not only in the commercial world, but also in battlefields and military settings, although this might be controversial. This is why Google decided not to renew their US government contract to develop AI techniques that can be used for military applications, such as battlefields (e.g., AI in Internet of military/battlefield things applications).

AI can also play an important role in cybersecurity, cyberthreat intelligence, and analytics to detect, contain, and mitigate advanced persistent threats (APTs), and fight against and mitigate malicious cyberactivities (e.g., organized cybercrimes and state-sponsored cyberthreats). For example, AI techniques can be trained to automatically scan for unknown malware or zero-day exploitation, based on certain features and behavior, rather than specific signatures.

This is the focus of this book. In Chap. 2, for example, the authors explain the potential of formal knowledge representation of network semantics in addressing interoperability challenges due to heterogeneous network data obtained from a wide range of sources. This constitutes a significant amount of data collected from or generated by, say, different security monitoring solutions. The formal knowledge representation of network semantics can be leveraged by intelligent and next-generation AI techniques and solutions to facilitate mining, interpreting, and extracting knowledge from structured (big) data to inform decision-making and cyberdefense strategies. Other cybersecurity applications of AI include the identification of vulnerabilities and weaknesses in systems and devices, and the detection of suspicious behaviors and anomalies, as explained in this book.

AI techniques and tools can, however, be exploited for malicious purposes. For example, an adversary can also use AI techniques to identify and exploit vulnerabilities in systems and devices. One possible scenario is for an attacker (or group of attackers) to design AI techniques to identify and exploit vulnerabilities in, say, driverless vehicles and unmanned aerial vehicles (UAVs, also known as drones), in order to facilitate coordinated attacks at places of mass gatherings (e.g., financial districts in cities during peak hours). Also, via coordinated attacks, AI techniques can exploit vulnerabilities in smart city infrastructures (e.g., intelligent transportation systems) to maximize the impact of such attacks, with the aim of causing societal panic and unrest. Hence, there is also a need to defend against AI-based cyberphysical attacks.

In other words, more work needs to be done in several directions to answer these open questions:

1. How to use AI techniques to facilitate forensic investigation?
2. How can we design AI techniques to facilitate the prediction of future cyberattacks or potential vulnerabilities (and their risk of successful exploitation)?
3. Can AI techniques help design new security solutions to overcome the pitfalls of human design (e.g., mitigate the "break-and-fix" trend in cryptographic protocols) or even design new blockchain types?

In short, the authors provide a comprehensive treatment of AI in cybersecurity, making this book useful to anyone interested in advancing the field of AI in cybersecurity.

Unquestionably, this is an exciting era, and the nexus between AI and cybersecurity will be increasingly important in the foreseeable future.

San Antonio, TX, USA Kim-Kwang Raymond Choo, Ph.D.

Preface

Security breaches and compromised computer systems cause significant losses for governments and enterprises alike. Attack mechanisms evolve in parallel with defense mechanisms, and detecting fraudulent payment gateways, protecting cloud services, and securely transferring files require next-generation techniques that are constantly improved. Artificial intelligence methods can be employed to fight against cyberthreats and attacks, with the aim of preventing as many future cyber-attacks as possible or at least minimizing their impact. The constantly increasing volume of global cyberthreats and cyberattacks urges automated mechanisms that are capable of identifying vulnerabilities, threats, and malicious activities in a timely manner. Knowledge representation and reasoning, automated planning, and machine learning are just some of those areas through which machines can contribute to proactive, rather than reactive, cybersecurity measures. To this end, AI-powered cybersecurity applications are developed—see the AI-based security infrastructure *Chronicle* and the "enterprise immune system" *Darktrace*, for example.

The increasing complexity of networks, operating system and wireless network vulnerabilities, and malware behavior poses a real challenge at both the national and international levels to security specialists. This book is a collection of state-of-the-art AI-powered security techniques and approaches. Chapter 1 introduces the reader to ontology engineering and its use in cybersecurity, cyberthreat intelligence, and cybersituational awareness applications. The formal definition of concepts, properties, relationships, and entities of network infrastructures in ontologies makes it possible to write machine-interpretable statements. These statements can be used for efficient indexing, querying, and reasoning over expert knowledge. Chapter 2 details how to employ knowledge engineering in describing network topologies and traffic flow so that software agents can process them automatically, and perform network knowledge discovery. Network analysts use information derived from diverse sources that cannot be efficiently processed by software agents unless a uniform syntax is used, associated with well-understood meaning fixed via interpretations, such as model-theoretic semantics and

Tarski-style interpretation. The formal representation of network knowledge requires not only crisp, but also fuzzy and probabilistic axioms, and metadata such as provenance. Reasoning over semantically enriched network knowledge enables automated mechanisms to generate non-trivial statements about correlations even experienced analysts would overlook, and automatically identify misconfigurations that could potentially lead to vulnerabilities.

Chapter 3 warns us that AI can not only assist, but might also pose a threat to cybersecurity, because hackers can also use AI methods, such as to attack machine learning systems to exploit software vulnerabilities and compromise program code. For example, they might introduce misleading data in machine learning algorithms (*data poisoning*), which can result in email campaigns that mark thousands of spam emails as "not spam" to skew the algorithm's behavior, thereby achieving malicious emails to be considered solicited.

Software vulnerabilities mentioned in social media sites can be used by software vendors for prompt patching, but also by adversaries to exploit them before they are patched. Using their hands-on skills, the authors of Chap. 4 demonstrate the correlation between publishing and exploiting software vulnerabilities, and propose an approach to predict the likelihood of vulnerability exploitation based on online resources, including the Dark Web and the Deep Web, which can be used by vendors for prioritizing patches.

Chapter 5 describes AI methods suitable for detecting network attacks. It presents a binary classifier and optimization techniques to increase classification accuracy, training speed, and the efficiency of distributed AI-powered computations for network attack detection. It describes models based on neural, fuzzy, and evolutionary computations, and various schemes for combining binary classifiers to enable training on different subsamples.

Chapter 6 discusses intrusion detection techniques that utilize machine learning and data mining to identify malicious connections. Common intrusion detection systems using fuzzy logic and artificial neural networks are critically reviewed and compared. Based on this comparison, the primary challenges and opportunities of handling cyberattacks using artificial neural networks are summarized.

Security is only as strong as the weakest link in a system, which can be easily demonstrated by the example of a careless or inexperienced user, who can compromise the security of a file transfer or a login regardless of enterprise security measures. For example, software installations may not only result in malware infections, but also in data loss, privacy breach, etc. Artificial intelligence can be utilized in the analysis of the security of software installers, as seen in Chap. 7, which investigates the machine learning-based analysis of the Android application package file format used on Android mobile devices for the distribution and installation of mobile apps and middleware. The analysis of the APK file structure makes it possible to understand those properties that can be used by machine learning algorithms to identify potential malware targeting.

Network analysts, defense scientists, students, and cybersecurity researchers can all benefit from the range of AI approaches compiled in this book, which not only reviews the state of the art, but also suggests new directions in this rapidly growing research field.

Adelaide, SA, Australia

Leslie F. Sikos, Ph.D.

Contents

About the Editor

Leslie F. Sikos, Ph.D., is a computer scientist specializing in formal knowledge representation, ontology engineering, and automated reasoning applied to various domains, including cyberthreat intelligence and network applications that require cybersituational awareness. He has worked in both academia and the enterprise, and acquired hands-on skills in datacenter and cloud infrastructures, cyberthreat management, and firewall configuration. He holds professional certificates and is a member of industry-leading organizations, such as the ACM, the Association for Automated Reasoning, the IEEE Special Interest Group on Big Data for Cyber Security and Privacy, and the IEEE Computer Society Technical Committee on Security and Privacy.

Chapter 1
OWL Ontologies in Cybersecurity: Conceptual Modeling of Cyber-Knowledge

Leslie F. Sikos

Abstract Network vulnerability checking, automated cyberthreat intelligence, and real-time cybersituational awareness require task automation that benefit from formally described conceptual models. Knowledge organization systems, including controlled vocabularies, taxonomies, and ontologies, can provide the network semantics needed to turn raw network data into valuable information for cybersecurity specialists. The formal knowledge representation of cyberspace concepts and properties in the form of upper and domain ontologies that capture the semantics of network topologies and devices, information flow, vulnerabilities, and cyberthreats can be used for application-specific, situation-aware querying and knowledge discovery via automated reasoning. The corresponding structured data can be used for network monitoring, cybersituational awareness, anomaly detection, vulnerability assessment, and cybersecurity countermeasures.

1.1 Introduction to Knowledge Engineering in Cybersecurity

Formal knowledge representation is a field of artificial intelligence, which captures the *semantics* (meaning) of concepts, properties, relationships, and individuals of specific *knowledge domains*, i.e., fields of interest or areas of concern, as structured data. The *Resource Description Framework (RDF)*[1] makes it possible to write machine-interpretable statements in the form of subject–predicate–object triples, called *RDF triples* [1]. These can be expressed using various syntaxes, among which

[1] https://www.w3.org/RDF/

L. F. Sikos (✉)
University of South Australia, Adelaide, SA, Australia
e-mail: leslie.sikos@unisa.edu.au

© Springer Nature Switzerland AG 2019

L. F. Sikos (ed.), *AI in Cybersecurity*, Intelligent Systems Reference Library 151,
https://doi.org/10.1007/978-3-319-98842-9_1

some of the most common ones are RDF/XML,[2] Turtle,[3] N-Triples,[4] JSON-LD,[5] RDFa,[6] and HTML5 Microdata.[7] In the Turtle serialization format, for example, the natural language sentence "WannaCry is a ransomware" can be written as the triple `:WannaCry rdf:type :Ransomware .`, where `type` is the `isA` relationship from the RDF vocabulary (rdfV) at http://www.w3.org/1999/02/22-rdf-syntax-ns#type.[8] Note that RDF triples are terminated by a period in Turtle, unless there are multiple statements with the same subject and predicate, but different objects, in which case the subject and the predicate are written just once, followed by the list of objects, separated by semicolons. Using the namespace mechanism, prefixes (such as `rdf:` in this example) can be abbreviated by defining the prefix in the beginning of the Turtle file as shown in Listing 1.1.

Listing 1.1 Namespace prefix declaration

```
@prefix rdf: <http://www.w3.org/1999/02/22-rdf-syntax-ns#> .
```

Such RDF triples are machine-interpretable, and using strict rules they can be used to infer new statements based on explicitly stated ones via *automated reasoning*.

Machine-interpretable definitions are created on different levels of conceptualization and comprehensiveness, resulting in different types of *knowledge organization systems* (KOS). *Controlled vocabularies* organize terms bound to preselected schemes for subsequent retrieval in subject indexing, describe subject headings, etc. (see Definition 1.1).

Definition 1.1 (Controlled Vocabulary) A *controlled vocabulary* is a triple $V = (\mathrm{N}_C, \mathrm{N}_R, \mathrm{N}_I)$ of countably infinite sets of IRI[9] symbols denoting atomic concepts (concept names or classes) (N_C), atomic roles (role names, properties, or predicates) (N_R), and individual names (objects) (N_I), respectively, where N_C, N_R, and N_I are pairwise disjoint sets.

Thesauri are reference works that list words grouped together according to similarity of meaning, such synonyms. *Taxonomies* are categorized words in hierarchy. *Knowledge bases* (KB) describe the terminology and individuals of a knowledge domain, including the relationships between concepts and individuals. *Ontologies* are formal conceptualizations of a knowledge domain with complex relationships and complex rules suitable for inferring new statements automatically. *Datasets* are collections of related machine-interpretable statements.

---

[2]https://www.w3.org/TR/rdf-syntax-grammar/

[3]https://www.w3.org/TR/turtle/

[4]https://www.w3.org/TR/n-triples/

[5]https://www.w3.org/TR/json-ld/

[6]https://www.w3.org/TR/rdfa-primer/

[7]https://www.w3.org/TR/microdata/

[8]Because this is a very common predicate, Turtle allows the abbreviation of `rdf:type` simply as `a`.

[9]Internationalized resource identifier.

The RDF vocabulary, which is capable of expressing core relationships, was extended to be able to describe more sophisticated relationships between concepts and properties, such as taxonomical structures, resulting in the *RDF Schema Language (RDFS)*.[10] Vocabularies, taxonomies, thesauri, and simple ontologies are usually defined in RDFS. Complex knowledge domains require even more representational capabilities, such as property cardinality constraints, domain and range restrictions, and enumerated classes, which led to the Web Ontology Language (OWL, purposefully abbreviated with the W and the O swapped),[11] a language specially designed for creating web ontologies with a rich set of modeling constructors and addressing the ontology engineering limitations of RDFS.

Each OWL ontology consists of RDF triples that define concepts (classes), roles (properties and relationships), and individuals.[12] The ontology file of an OWL ontology can be defined by the ontology URI, the `rdf:type` predicate, and the `Ontology` concept from the OWL vocabulary (see Listing 1.2).

Listing 1.2 An OWL file defined as an ontology

```
@prefix rdf: <http://www.w3.org/1999/02/22-rdf-syntax-ns#> .
@prefix owl: <http://www.w3.org/2002/07/owl#> .
<http://example.com/cyberontology.owl> a owl:Ontology .
```

Classes can be declared using `rdf:type` and the `Class` concept from the OWL vocabulary as shown in Listing 1.3.

Listing 1.3 An OWL class declaration

```
:Malware a owl:Class .
```

Superclass-subclass relationships can be defined using the `subclassOf` property from the RDFS vocabulary as shown in Listing 1.4.

Listing 1.4 An OWL file defined as an ontology

```
:Ransomware rdfs:subClassOf :Malware .
```

In OWL, properties that define relationships between entities are declared using `owl:ObjectProperty`, and those that assign properties to property values or value ranges with `owl:DatatypeProperty`, as demonstrated in the examples in Listing 1.5.

Listing 1.5 An object property and a datatype property declaration

```
:connectedTo a owl:ObjectProperty .
:hasIPAddress a owl:DatatypeProperty .
```

Note that OWL concept names are typically written in PascalCase, in which words are created by concatenating capitalized words, and OWL property names are usually written in camelCase, i.e., compound words or phrases are written in a way that each word or abbreviation begins with a capital letter (medial capitalization).

Individuals can be declared using the `namedIndividual` term from the OWL vocabulary as shown in Listing 1.6.

[10] https://www.w3.org/TR/rdf-schema/

[11] https://www.w3.org/OWL/

[12] These may be complemented by SWRL rules, although doing so can result in undecidability.

Listing 1.6 An entity declaration

```
:WannaCry a owl:namedIndividual .
```

Individuals can also be declared as an instance of a class, as seen in our very first example. In OWL, many other constructors are available as well, such as atomic and complex concept negation, concept intersection, universal restrictions, limited existential quantification, transitivity, role hierarchies, inverse roles, functional properties, datatypes, nominals, and cardinality restrictions. In the second version of OWL, OWL 2, these are complemented by the union of a finite set of complex role inclusions and a role hierarchy, and unqualified cardinality restrictions are replaced by qualified cardinality restrictions.[13]

Formal grounding for OWL ontologies can be provided by description logics, many of which are decidable fragments of first-order predicate logic and feature a different balance between expressivity and reasoning complexity by supporting different sets of mathematical constructors [2].

The uniform representation of expert cyber-knowledge enables important data validation tasks, such as automated data integrity checking, and inference mechanisms to make implicit statements explicit via automated reasoning.

The scope of an ontology is determined by granularity. *Upper ontologies*, also known as upper-level ontologies, top-level ontologies, or foundational ontologies, are generic ontologies applicable to various domains. *Domain ontologies* describe the vocabulary of a specific knowledge domain with a specific viewpoint and factual data. *Core reference ontologies* are standardized or de facto standard ontologies used by different user groups to integrate their different viewpoints about a knowledge domain by merging several domain ontologies.

1.2 Cybersecurity Taxonomies

The *Taxonomy of Dependable and Secure Computing* defines a concept hierarchy for attributes (availability, reliability, safety, confidentiality, integrity, maintainability), threats (faults, errors, failures), and means (fault prevention, fault tolerance, fault removal, fault forecasting) [3]. The taxonomical structure details these concepts further. Faults are classified as development faults, physical faults, and interaction faults. Failures can be service failures, development failures, or dependability failures. Fault tolerance has subclasses such as error detection and recovery. Fault removal defines more specific classes, such as verification, diagnosis, correction, and non-regression verification. Fault forecasting is either ordinal evaluation or probabilistic evaluation, the latter of which can be modeling or operational testing. The fault classes are particularly detailed in a very deep class hierarchy.

Hansman and Hunt created a taxonomy of network and computer attacks to categorize attack types [4]. It consists of four "dimensions," which categorize attacks

[13]Providing examples for each constructor is beyond the scope of this chapter. For a detailed description, see https://www.w3.org/TR/owl2-quick-reference/

by attack classes, attack targets, vulnerabilities and exploits used by the attack, and the possibility whether the attack has a payload or effect beyond itself. To achieve this, the first dimension defines concepts such as virus, worm, Trojan, buffer overflow, denial of service attack, network attack, physical attack, password attack, and information gathering attacks. The second dimension has both hardware and software categories, including computer, operating system, application, and network, all detailed to be able to express the highest level of granularity, such as operating system versions and editions, and networking protocols. The third dimension reuses de facto standard *Common Vulnerabilities and Exposures (CVE)*[14] entries whenever possible. The fourth dimension describes possible payloads, such as a worm may have a Trojan payload.

The attack taxonomy of Gao et al. categorizes attacks as follows [5]:

- Attack impact: confidentiality, integrity, availability, authentication, authorization, auditing
- Attack vector: DoS, human behavioral attack, information gathering, malformed input, malicious code, network attack, password attack, physical attack
- Attack target: hardware (computer, network equipment, peripheral device), software (application, network, operating system), human (receiver, sender)
- Vulnerability: location, motivation, resource-specific weaknesses, weaknesses introduced during design, weaknesses introduced during implementation, weaknesses in OWASP[15] top ten
- Defense: security network communication, security hardware, standard, emissions security, technology mistakenly used as countermeasure, backup, encryption, cryptography, message digest checksum, memory protection, trust management, access control, vulnerability scanner, source code scanner, login system, monitoring, honeypot, sanitizer, key management

This taxonomy was designed to describe attack effect evaluation by capturing security property weights and evaluation indices, where the attack effect covers system performance changes both before and after attacks.

The taxonomy of cyberthreat intelligence information exchange of Burger et al. is aligned with OpenIOC,[16] the Structured Threat Information Expression (STIX),[17] and the Incident Object Description Exchange Format (IODEF)[18] [6]. In contrast to taxonomies with a class hierarchy, it adopts a layered model, which partially follows the ISO/OSI model. The five layers from bottom to top are transport, session, indicators, intelligence, and 5W's. The transport conceptual layer contains synchronous byte streams, asynchronous atomic messages, and raw byte streams. The session layer covers authenticated senders and receivers, and permissions on entire content. The indicators layer defines patterns, behaviors, and permission on indicator.

[14]https://cve.mitre.org

[15]https://www.owasp.org

[16]http://www.openioc.org

[17]https://oasis-open.github.io/cti-documentation/stix/intro

[18]https://www.ietf.org/rfc/rfc5070.txt

Intelligence covers three concepts: action, query, and target. The top layer contains the 5W's and the H (Who, What, When, Where, Why, How).

1.3 A Core Reference Ontology for Cybersecurity

The *Reference Ontology for Cybersecurity Operational Information* is a cybersecurity ontology that structures cybersecurity information and orchestrates industry specifications from the viewpoint of cybersecurity operations [7]. It is aimed at serving as the basis for cybersecurity information exchange on a global scale using standards from ISO/IEC,[19] OASIS,[20] NIST,[21] ITU-T,[22] MITRE,[23] the Open Grid Forum,[24] and IEEE.[25] The implementation of this ontology allows vulnerabilities to be described using de facto standard CVE identifiers and threats using *Common Attack Pattern Enumeration and Classification (CAPEC)* identifiers.[26] The Reference Ontology for Cybersecurity Operational Information is aligned with standard guidelines such as ISO/IEC 27032,[27] ITU-T E.409,[28] ITU-T X.1500,[29] and IETF RFC 2350.[30]

1.4 Upper Ontologies for Cybersecurity

The *Security Ontology (SO)* was developed to be used for representing arbitrary information systems [8]. It can model assets such as data, network, and services, countermeasures such as firewalls and antivirus software, and risk assessment knowledge. The Security Ontology can be used to formalize security requirements, and supports the aggregation of, and reasoning about, interoperable security knowledge derived from diverse sources.

The *Security Asset Vulnerability Ontology (SAVO)* is an upper ontology for information security, and was specially designed to capture core concepts such as threat, risk, DoS attack, illegal access, and vulnerability, and the properties of these

[19] https://www.iso.org

[20] https://www.oasis-open.org

[21] https://www.nist.gov

[22] https://www.itu.int

[23] https://www.mitre.org

[24] https://www.ogf.org

[25] https://www.ieee.org

[26] https://capec.mitre.org

[27] https://www.iso.org/standard/44375.html

[28] https://www.itu.int/rec/T-REC-E.409-200405-I/en

[29] https://www.itu.int/rec/T-REC-X.1500/en

[30] https://www.ietf.org/rfc/rfc2350.txt

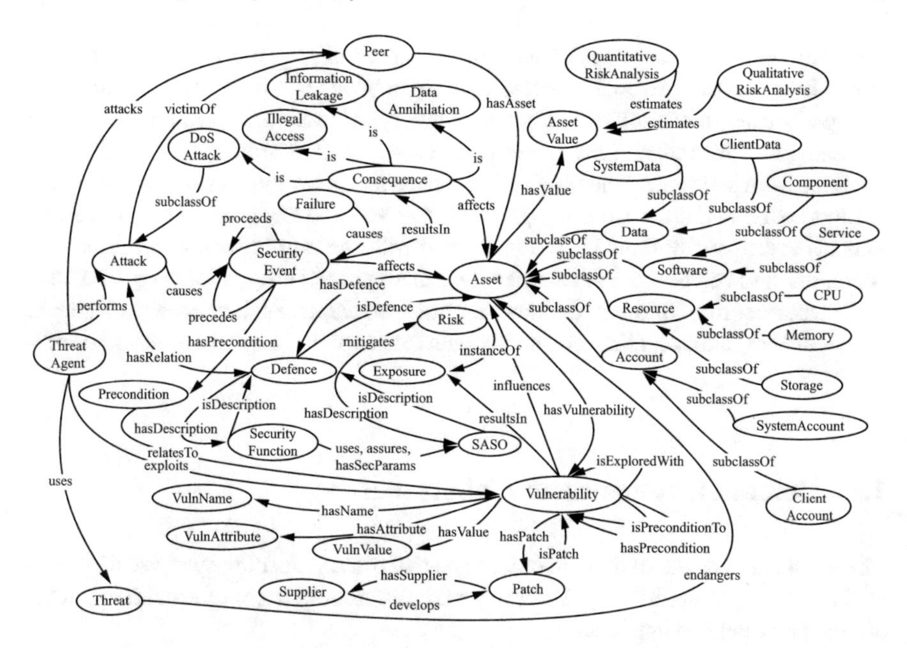

Fig. 1.1 Structure of the Security Asset Vulnerability Ontology [9]

concepts. It defines the relationship between threat, vulnerability, risk, exposure, attack, and other security concepts, and maps low-level and high-level security requirements and capabilities (see Fig. 1.1).

The *Security Algorithm-Standard Ontology (SASO)*, developed by the same authors as SAVO, encompasses security algorithms, standards, concepts, credentials, objectives, assurance levels. It consists of concepts such as security algorithm, security assurance, security credential, and security objective. The security ontology of Fenz et al. was created to capture the semantics of risk management, covering security concepts such as attributes, vulnerabilities, threats, controls, ratings, and alike [10]. The attribute type contains subtypes such as control type, threat origin, scale, etc. The control concept covers the control type, relation to established information security standards, implementation specification, and mitigation specification. Vulnerability covers both administrative vulnerabilities and technical vulnerabilities. The definitions in the ontology are formally grounded with description logic axioms, and are aligned with standards such as ISO 27001[31] and the NIST information security risk management guide [11]. Wali et al. generated a cybersecurity ontology from textbook index terms via automated classification using bootstrapping strategies [12]. The ontology is rather comprehensive, with a deep taxonomical structure. For the `Countermeasure` concept, for example, it defines subconcepts such as `AccessControl`, `Backup`, `Checksum`, `Cryptography`, `EmissionsSecurity`, `HoneyPot`, and `KeyManagement`.

[31] https://www.iso.org/isoiec-27001-information-security.html

The ontology engineers considered terminological differences and alternate categories by using two textbook indices, along with Wikipedia categories. Because the Wikipedia categories provided either no category information or too many ambiguous categories for index terms, however, the ontology is dominated by terminology obtained from textbooks. The *Unified Cybersecurity Ontology (UCO)*[32] was designed to incorporate and integrate heterogeneous data and knowledge schemata from various cybersecurity systems [13]. It is aligned with the most common cybersecurity standards, and can be used to complement other cybersecurity ontologies. The Unified Cybersecurity Ontology defines important concepts, such as means of attack, consequences, attacks, attackers, attack patterns, exploits, exploit targets, and indicators.

1.5 Domain Ontologies for Cybersecurity

There is an abundance of domain ontologies that provide specific cybersecurity concept and property definitions. All of these are different in terms of specificity, granularity, and intended application.

1.5.1 Intrusion Detection Ontologies

He et al. introduced an ontology for intrusion detection, which captures semantics for attack signatures including, but not limited to, system status (connection status, CPU and memory usage), IP address, port, protocol, and router and firewall logs [14]. Vulnerability representations in this ontology are aligned with common CVE codes. The ontology can be used for ontology-based signature intrusion detection using string matching, and for multi-sensor cooperative detection.

1.5.2 Malware Classification and Malware Behavior Ontologies

The *Cyber Ontology* of MITRE is based on a malware ontology, enables integration across disparate data sources, and supports automated cyber-defense tasks [15]. The conceptual framework behind this ontology is Ingle's diamond model of malicious activity, which includes actors, victims, infrastructure, and capabilities. The *Ontology for Malware Behavior* is an ontology based on a set of suspicious behaviors observed during malware infections on the victims' systems [16]. It defines the following types of concepts:

[32]https://github.com/Ebiquity/Unified-Cybersecurity-Ontology/blob/master/uco_1_5_rdf.owl

- Attack launch events, such as denial of service, email sending (spam, phising), scanning, exploit sending
- Evasion, including anti-analysis (debugger checking, environment detection, removal of evidence), anti-defense (removal of registries, shutdown of defense mechanisms)
- Remote control, such as download code (known malware execution, other code execution, get command)
- Self-defense, including maintenance events, such as component checking, creating synchronization object, language checking, persistence
- Stealing, including system information stealing (hostname, OS information, resource information) and user information stealing (credential, Internet banking data)
- Subversion, such as browser, memory writing, and operating system

The ontology can be used to describe malware actions with RDF statements. Swimmer proposed *malonto*, a malware ontology to be used for hypothesis testing and root cause analysis via reasoning. It defines classes for viruses, worms, Trojans, exploits, droppers, and properties such as `drops`, `hasPayload`, `hasInsituacy`, `hasLocality`, `hasObfuscation`, `hasTransitivity`, and `hasPlurality`.

The *Malware Ontology*[33] is a comprehensive ontology containing malware categories, such as Trojan, exploit categories, such as XSS and SQLi, and about 12,000 malware names. This ontology was designed for threat intelligence teams to assist in completing Big Data tasks in a timely manner. It can be used to summarize malware targets, answer questions related to patterns, describe recently reported IPs, hashes, and file names, and aggregate and link technical information related to malware.

1.5.3 Ontologies for Cyberthreat Intelligence

There are ongoing community efforts to standardize cyberthreat information, as evidenced by the STIX and the IODEF formats, and ontologies that define terms aligned with them [17]. In fact, several cyberthreat ontologies and based on STIX concepts, and some extend STIX [18], although there are threat ontologies that are more specific or are not aligned with de facto standards.

The *Incidence Response and Threat Intelligence Ontology*[34] is, as its name suggests, an ontology for classifying and analyzing Internet entities with an emphasis on computer incident response and threat intelligence processing. It defines semantics for data formats used in this field, making it unnecessary to re-interpret data in case it is not trivial how the data was produced.

Ekelhart et al. created an ontology based on the Taxonomy of Dependable and Secure Computing mentioned in Sect. 1.2 and concepts of the IT infrastructure

[33] https://www.recordedfuture.com/malware-ontology/

[34] https://raw.githubusercontent.com/mswimmer/IRTI-Ontology/master/irti.rdf

domain [19]. It can be used to formally describe the allocation of relevant IT infrastructure elements in a building, including floor and room numbers. Such descriptions can be used for defining disasters, such as a physical threat. When used for simulation, the ontology can provide the basis for calculating IT costs, analyze the impact of a particular threat, potential countermeasures, and their benefits.

The insider threat indicator ontology of Costa et al.[35] defines concepts for describing potential indicators of malicious insider activity, insider threat detection, prevention, and mitigation [20]. The ontology supports the creation, sharing, and analysis of indicators of insider threats with concepts such as `AcceptAction`, `AccessAction`, `AccountAuthenticationInformation`, `AnomalousEvent`, `BackdoorSoftwareAsset`, `BankAccountAsset`, `BreachAction`, `ClassifiedInformation`, `DataDeletionEvent`, `DataExfiltrationEvent`, `DataModificationEvent`, `EncryptAction`, `FailedAction`, `FraudulentAction`, `IllegitimateAction`, and `IntellectualProperty`.

The *Ontology for Threat Intelligence* is an ontology that implements the widely adopted intrusion kill chain of Lockheed-Martin [21], capable of describing seven attack stages, each of which is associated with controls through which intrusion can be disrupted, namely reconnaissance, weaponization, delivery, exploit, install, command and control, and actions on target [22]. The ontology is designed to help threat intelligence analysts effectively organize and search open source intelligence and threat indicators, thereby acquiring a better understanding of the threat environment. The Ontology for Threat Intelligence can potentially be used to describe various attacks, such as distributed denial of service (DDoS) attacks, which generate an incoming traffic that floods the victim from multiple unique IP addresses, and industrial control system attacks, which may result in power outages in the energy sector.

1.5.4 The Ontology for Digital Forensics

The *Ontology for Digital Forensics in IT Security Incidents* defines all forensically relevant parts of computers for a particular point in time [23]. Each forensic object is associated with a timestamp and the forensic tool by which it was retrieved. In terms of hardware, these include concepts such as memory, HDD, network interface card (NIC). As for software, the ontology defines the kernel, resources, and process list concepts, which are used in relation definitions such as `hasResource` and `processlist`. The user concept is characterized by properties such as name and password, and one or more groups the user belongs to. The ontology also defines the process concept, which is associated with every program, and all processes are listed in the process list. By definition, each of them, except the initial one, has at least one thread and one parent process. Networking terms are defined as NIC

[35]http://resources.sei.cmu.edu/asset_files/TechnicalReport/2016_005_112_465537.owl

configurations, which include IP, gateway, and name server. The connections, which are wrapped by sockets, feature local and remote IP addresses. The port and protocol defined for sockets can be used by processes to reference resources. The registry concept contain hives, each of which has a root key. Each key has a name and a state that represent the flag where the key can be found. Keys may have optional subkeys and values. Values contain key-value pairs, a value type, and a state. For file systems, the ontology defines file system categories, data unit categories, file name categories, metadata categories, and application categories. The Memory concept has five subclasses: memory system architecture, metadata, metacode, data, and code. Metadata has more specific subclasses, namely, memory organization metadata and runtime organization metadata. The subclasses of the data concept are OS-specific data and application data, and the subclasses of the code concept are OS-specific code and application code.

1.5.5 Ontologies for Secure Operations and Processes

Oltramari et al. created a formally grounded ontology for secure operations [24]. Among other things, it can be used to describe the task of retrieving files securely and detect intrusions. It features concepts such as cyber-operation, cyber-operator, mission plan, cyber-exploitation, payload, etc., and the relationships between these concepts.

Maines et al. created an ontology for the cybersecurity requirements of business processes [25]. It defines concepts at four levels, the highest of which contains the key concepts: privacy, access control, availability, integrity, accountability, and attack/harm detection and prevention. Privacy has two direct subclasses: confidentiality and user consent. Access control defines subclasses such as authentication, identification, and authorization. Availability includes service, data, personnel, and hardware backups. Integrity covers data, hardware, personnel, and software integrity control. The accountability class has subclasses such as audit trail, digital forensics, and non-repudiation. Attack/harm detection and prevention includes the following concepts: vulnerability assessment, honeypot, firewall, and intrusion detection and prevention system. These concepts are detailed further on the third and fourth level.

1.5.6 An Ontology for Describing Cyberattacks and Their Impact

The *Cyber Effects Simulation Ontology (CESO)*[36] is suitable for describing cyber-impacts in terms of not only qualitative outcomes, but also technical and human

[36]https://github.com/AustralianCentreforCyberSecurity/Cyber-Simulation-Terrain/blob/master/v1-0/cyber/ThreatSimulationOntology/tso.ttl

aspects [26]. It is aligned with STIX, and defines the classes campaign, course of action, exploit, exploit target, and threat actor, and their properties. The ontology was designed for modeling and simulating the effects of cyberattacks on organizations and military units, such as an artillery fire mission in a land combat. To do so, it not only models virtual concepts, but also physical concepts and events.

1.6 Networking Ontologies for Cybersecurity

The formal representation of network infrastructures provides standardized data schemas, which may enable task automation via machine learning for vulnerability and exposure analysis [27]. These infrastructures include not only computers, but also Internet of Things (IoT) devices, which can be described efficiently using ontologies [28]. Ontology-based representations can also be used for infrastructureless wireless networks, such as mobile ad hoc networks (MANET),[37] which can be described using the *MANET Distributed Functions Ontology (MDFO)* [29].

The first network management ontologies date back to the early days of the Semantic Web (see [30], for example). The network ontology of De Paola et al. for computer network management defines concepts such as traffic entity, event, database, management tool, traffic statistics, action, demand, actor, routing, abnormality, and net entity [31]. The *Network Domain Ontology* was designed for detailed network representations, where network devices are instances of network concepts, which can be described using properties such as MAC address, IP address, operating system, and the number of the office in which the device is located [32].

The *Network Ontology (NetOnto)* is an ontology designed to describe exchanged BGP[38] routes [33]. The OWL axioms of the ontology are complemented by BGP policies written as SWRL[39] rules, which determine how routing information is shared between neighbors to control traffic flow across networks. These policies include contextual information such as affiliations and route restrictions as well. Using NetOnto allows focusing on high-level policies and reconfigure networks automatically.

The ontology of Basile et al. defines concepts at three levels of granularity: first-level (e.g., `Switch`, `Router`, `Firewall`), second-level (`Workstation`, `Server`, `PublicService`, etc.), and third-level concepts (e.g., `SharedWorkstation`, `PublicServer`), out of which the letter two can be automatically generated via reasoning over the first [34].

The firewall ontology of Ghiran et al. defines concepts for ranges of ports, ranges of IP addresses, interfaces, firewall rules, protocols, and actions [35]. The firewall rules are further classified as correct rules, conflicting rules, and unclassified rules. Conflicting rules are defined to be generalizations, redundant rules, shadowed rules,

---

[37] https://tools.ietf.org/html/rfc2501

[38] Border Gateway Protocol.

[39] Semantic Web Rule Language.

or correlated rules. The ontology is best complemented by SWRL rules for advanced reasoning over firewall policies.

The *SCADA Ontology* of the INSPIRE project defines asset, vulnerability, treat, attack source, and safeguard concepts based on the now-withdrawn ISO/IEC 13335-1:2004 standard,[40] and the attributes, relations, and interdependencies of these concepts [36]. It captures the semantics of the connections and dependencies among SCADA systems and their security aspects.

The *Physical Devices Ontology* defines concepts such as device, card, configuration, and slot, along with their object and datatype properties [37]. It can be used to describe devices, such as routers, and their configuration.

The *Measurement Ontology for IP Traffic* is an ETSI specification for interfacing and data exchange with IP traffic measurement devices [38]. It merges ontologies that define concepts such as connection, protocol, application and routing device, logistic and physical location, routing and queueing algorithm, and units for measuring network traffic. It is aligned with standard information models for network measurements, such as IETF SNMP MIBs (RFC 3418),[41] IPFIX (RFC 3955),[42] and IPPM (RFC 6576),[43] CAIDA DatCat,[44] ITU M.3100,[45] and DMTF CIM.[46]

The *Ontology for the Router Configuration Domain (ORCONF)* can be used to describe the semantics of the command-line interface (CLI) of network routers [39]. More specifically, it is suitable for the description of all executable configuration statements of routers, i.e., valid sequences of commands and variables used in combination to represent a router configuration operation (e.g., set → system → host_name → <name>). Other devices, such as switches, can be described similarly using the *Ontology for Network Device Configuration (ONDC)*, which can be used as part of a framework for populating vendor- and device-specific ontologies [40].

The probabilistic ontology of Laskey et al. was designed for large-scale IP address geolocation [41]. It integrates a factor graph model for IP geolocation with a domain ontology representing geolocation knowledge. Random variables in the factor graph model correspond to uncertain properties and relationships in the domain ontology. The model can be used to reason over arbitrary numbers of IP nodes, regions, and network topologies.

MonONTO is a domain ontology for network monitoring [42]. It defines concepts for the quality of service of advanced applications, network performance measurements, and user profiles. It can be used to monitor the performance of advanced Internet applications based on the use of an expert system. MonONTO can capture the semantics of latencies, bandwidth throughput, link utilization, traceroutes, network losses, and delays. The developers of the ontology also created a set of

[40] https://www.iso.org/standard/39066.html

[41] https://tools.ietf.org/html/rfc3418

[42] https://www.ietf.org/rfc/rfc3955.txt

[43] https://tools.ietf.org/html/rfc6576

[44] http://www.datcat.org

[45] https://www.itu.int/rec/T-REC-M.3100/en

[46] https://www.dmtf.org/standards/cim

inference rules that consist of network knowledge derived from the relationships between MonONTO concepts.

The *IP Networks Topology and Communications Ontology*[47] is one of the most comprehensive ontologies that defines IPv4 and IPv6 network topology and connection concepts. It also defines custom datatypes for HTTP methods, IP addresses, protocols, and connection states.

The *Communication Network Topology and Forwarding Ontology (CNTFO)*[48] is based on the *Internet Protocol (IP) Ontology*, the *Open Shortest Path First (OSPF) Ontology*, and the *Border Gateway Protocol (BGP) Ontology* [43]. It captures the semantics of routing information retrieved from OSPF links state advertisements and BGP update messages, router configuration files, and open datasets. In addition, it covers fundamental administrative and networking concepts and their properties for describing autonomous systems, network topologies, and traffic flow. The property value ranges of the specific properties in the ontology are aligned with standard definitions from IETF RFCs and defined using restrictions on standard XSD datatypes. To make automatically generated data authoritative, the Communication Network Topology and Forwarding Ontology supports provenance data to be attached to cyber-knowledge both at the RDF triple and at the dataset level using standard PROV-O terms[49] and custom, very specific provenance terms, such as `importHost`, `importUser`, and `sourceType`. These provenance-aware statements can be captured using RDF quadruples, such as using the TriG[50] serialization format [44].

1.7 Summary

Ontology engineering can provide solutions to cybersecurity challenges via supporting automated data processing, which requires a common form of representation for expert cyber-knowledge. Considering the characteristics of RDF triples, using OWL ontologies for defining the semantics of the concepts and roles used in RDF triples describing cyber-knowledge is trivial. This is witnessed by the proliferation of cybersecurity ontologies ranging from upper ontologies to very specific domain ontologies, which can be used in combination to describe complex statements about IP infrastructures, network topology, current network status, and potential cyberthreats alike. Based on these formal descriptions, automated tasks can be performed efficiently, and non-trivial, implicit statements can be made explicit, thereby facilitating knowledge discovery in the increasingly complex and highly dynamic cyberspace.

[47]https://github.com/twosixlabs/icas-ontology/blob/master/ontology/ipnet.ttl

[48]http://purl.org/ontology/network/

[49]https://www.w3.org/TR/prov-o/

[50]https://www.w3.org/TR/trig/

References

1. Sikos LF (2015) Mastering structured data on the Semantic Web: from HTML5 Microdata to Linked Open Data. Apress, New York. https://doi.org/10.1007/978-1-4842-1049-9
2. Sikos LF (2017) Description logics in multimedia reasoning. Springer, Cham. https://doi.org/10.1007/978-3-319-54066-5
3. Avizienis A, Laprie J-C, Randell B, Landwehr C (2004) Basic concepts and taxonomy of dependable and secure computing. IEEE Trans Depend Secur Comput 1(1):11–33. https://doi.org/10.1109/TDSC.2004.2
4. Hansman S, Hunt R (2005) A taxonomy of network and computer attacks. Comput Secur 24(1):31–43. https://doi.org/10.1016/j.cose.2004.06.011
5. Gao J, Zhang B, Chen X, Luo Z (2013) Ontology-based model of network and computer attacks for security assessment. J Shanghai Jiaotong Univ (Sci) 18(5):554–562. https://doi.org/10.1007/s12204-013-1439-5
6. Burger EW, Goodman MD, Kampanakis P (2014) Taxonomy model for cyber threat intelligence information exchange technologies. In: Ahn G-J, Sander T (eds) Proceedings of the 2014 ACM Workshop on Information Sharing & Collaborative Security. ACM, New York, pp 51–60. https://doi.org/10.1145/2663876.2663883
7. Takahashi T, Kadobayashi Y (2015) Reference ontology for cybersecurity operational information. Comput J 58(10):2297–2312. https://doi.org/10.1093/comjnl/bxu101
8. Tsoumas B, Papagiannakopoulos P, Dritsas S, Gritzalis D (2006) Security-by-ontology: a knowledge-centric approach. In: Fischer-Hübner S, Rannenberg K, Yngström L, Lindskog S (eds) Security and privacy in dynamic environments. Springer, Boston, pp 99–110. https://doi.org/10.1007/0-387-33406-8_9
9. Vorobiev A, Bekmamedova N (2007) An ontological approach applied to information security and trust. In: Cater-Steel A, Roberts L, Toleman M (eds) ACIS2007 Toowoomba 5–7 December 2007: Delegate Handbook for the 18th Australasian Conference on Information Systems. University of Southern Queensland, Toowoomba, Australia. http://aisel.aisnet.org/acis2007/114/
10. Fenz S, Ekelhart A (2009) Formalizing information security knowledge. In: Li W, Susilo W, Tupakula U, Safavi-Naini R, Varadharajan V (eds) Proceedings of the 4th International Symposium on Information, Computer, and Communications Security. ACM, New York, pp 183–194. https://doi.org/10.1145/1533057.1533084
11. Stoneburner G, Goguen A, Feringa A (2002) Risk management guide for information technology systems. NIST Special Publication 800-30, National Institute of Standards and Technology (NIST), Gaithersburg, MD, USA
12. Wali A, Chun SA, Geller J (2013) A bootstrapping approach for developing a cyber-security ontology using textbook index terms. In: Guerrero JE (ed) Proceedings of the 2013 International Conference on Availability, Reliability, and Security. IEEE Computer Society, Washington, pp 569–576. https://doi.org/10.1109/ARES.2013.75
13. Syed Z, Padia A, Mathews ML, Finin T, Joshi A (2016) UCO: a unified cybersecurity ontology. In: Wong W-K, Lowd D (eds) Proceedings of the Thirtieth AAAI Workshop on Artificial Intelligence for Cyber Security. AAAI Press, Palo Alto, CA, USA, pp 195–202. https://www.aaai.org/ocs/index.php/WS/AAAIW16/paper/download/12574/12365
14. He Y, Chen W, Yang M, Peng W (2004) Ontology-based cooperative intrusion detection system. In: Jin H, Gao GR, Xu Z, Chen H (eds) Network and parallel computing. Springer, Heidelberg, pp 419–426. https://doi.org/10.1007/978-3-540-30141-7_59
15. Obrst L, Chase P, Markeloff R (2012) Developing an ontology of the cyber security domain. In: Costa PCG, Laskey KB (eds) Proceedings of the Seventh International Conference on Semantic Technologies for Intelligence, Defense, and Security. RWTH Aachen University, Aachen, pp 49–56. http://ceur-ws.org/Vol-966/STIDS2012_T06_ObrstEtAl_CyberOntology.pdf
16. Grégio A, Bonacin R, Nabuco O, Afonso VM, De Geus PL, Jino M (2014) Ontology for malware behavior: a core model proposal. In: Reddy SM (ed) Proceedings of the 2014 IEEE

23rd International WETICE Conference. IEEE, New York, pp 453–458. https://doi.org/10.1109/WETICE.2014.72
17. Asgarli E, Burger E (2016) Semantic ontologies for cyber threat sharing standards. In: Proceedings of the 2016 IEEE Symposium on Technologies for Homeland Security. IEEE, New York. https://doi.org/10.1109/THS.2016.7568896
18. Ussath M, Jaeger D, Cheng F, Meinel C (2016) Pushing the limits of cyber threat intelligence: extending STIX to support complex patterns. In: Latifi S (ed) Information technology: new generations. Springer, Cham, pp 213–225. https://doi.org/10.1007/978-3-319-32467-8_20
19. Ekelhart A, Fenz S, Klemen M, Weippl E (2007) Security ontologies: improving quantitative risk analysis. In: Sprague RH (ed) Proceedings of the 40th Annual Hawaii International Conference on System Sciences. IEEE Computer Society, Los Alamitos, CA, USA. https://doi.org/10.1109/HICSS.2007.478
20. Costa DL, Collins ML, Perl SJ, Albrethsen MJ, Silowash GJ, Spooner DL (2014) An ontology for insider threat indicators: development and applications. In: Laskey KB, Emmons I, Costa PCG (eds) Proceedings of the Ninth Conference on Semantic Technology for Intelligence, Defense, and Security. RWTH Aachen University, Aachen, pp 48–53. http://ceur-ws.org/Vol-1304/STIDS2014_T07_CostaEtAl.pdf
21. Falk C (2016) An ontology for threat intelligence. In: Koch R, Rodosek G (eds) Proceedings of the 15th European Conference on Cyber Warfare and Security. Curran Associates, Red Hook, NY, USA
22. Hutchins EM, Cloppert MJ, Amin RM (2011) Intelligence-driven computer network defense informed by analysis of adversary campaigns and intrusion kill chains. In: Armistead EL (ed) Proceedings of the 6th International Conference on Information Warfare and Security. Academic Conferences and Publishing International, Sonning Common, UK, pp 113–125
23. Wolf JP (2013) An ontology for digital forensics in IT security incidents. M.Sc. thesis, University of Augsburg, Augsburg, Germany
24. Oltramari A, Cranor LF, Walls RJ, McDaniel P (2014) Building an ontology of cyber security. In: Laskey KB, Emmons I, Costa PCG (eds) Proceedings of the Ninth Conference on Semantic Technology for Intelligence, Defense, and Security. RWTH Aachen University, Aachen, pp 54–61. http://ceur-ws.org/Vol-1304/STIDS2014_T08_OltramariEtAl.pdf
25. Maines CL, Llewellyn-Jones D, Tang S, Zhou B (2015) A cyber security ontology for BPMN-security extensions. In: Wu Y, Min G, Georgalas N, Hu J, Atzori L, Jin X, Jarvis S, Liu L, Calvo RA (eds) Proceedings of the 2015 IEEE International Conference on Computer and Information Technology; Ubiquitous Computing and Communications; Dependable, Autonomic and Secure Computing; Pervasive Intelligence and Computing. IEEE, New York, pp 1756–1763. https://doi.org/10.1109/CIT/IUCC/DASC/PICOM.2015.265
26. Ormrod D, Turnbull B, O'Sullivan K (2015) System of systems cyber effects simulation ontology. In: Proceedings of the 2015 Winter Simulation Conference. IEEE, New York, pp 2475–2486. https://doi.org/10.1109/WSC.2015.7408358
27. Sicilia MA, García-Barriocanal E, Bermejo-Higuera J, Sánchez-Alonso S (2015) What are information security ontologies useful for? In: Garoufallou E, Hartley R, Gaitanou P (eds) Metadata and semantics research. Springer, Cham, pp 51–61. https://doi.org/10.1007/978-3-319-24129-6_5
28. Gaglio S, Lo Re G (eds) (2014) Advances onto the Internet of Things: how ontologies make the Internet of Things meaningful. Springer, Cham. https://doi.org/10.1007/978-3-319-03992-3
29. Orwat ME, Levin TE, Irvine CE (2008) An ontological approach to secure MANET management. In: Jakoubi S, Tjoa S, Weippl ER (eds) Proceedings of the Third International Conference on Availability, Reliability and Security. IEEE Computer Society, Los Alamitos, CA, USA, pp 787–794. https://doi.org/10.1109/ARES.2008.183
30. De Vergara JEL, Villagra VA, Asensio JI, Berrocal J (2003) Ontologies: giving semantics to network management models. IEEE Netw 17(3):15–21. https://doi.org/10.1109/MNET.2003.1201472
31. De Paola A, Gatani L, Lo Re G, Pizzitola A, Urso A (2003) A network ontology for computer network management. Technical report No 22. Institute for High Performance Computing and Networking, Palermo, Italy

32. Abar S, Iwaya Y, Abe T, Kinoshita T (2006) Exploiting domain ontologies and intelligent agents: an automated network management support paradigm. In: Chong I, Kawahara K (eds) Information networking. Advances in data communications and wireless networks. Springer, Heidelberg, pp 823–832. https://doi.org/10.1007/11919568_82
33. Kodeswaran P, Kodeswaran SB, Joshi A, Perich F (2008) Utilizing semantic policies for managing BGP route dissemination. In: 2008 IEEE INFOCOM Workshops. IEEE, Piscataway, NJ, USA. https://doi.org/10.1109/INFOCOM.2008.4544611
34. Basile C, Lioy A, Scozzi S, Vallini M (2009) Ontology-based policy translation. In: Herrero Á, Gastaldo P, Zunino R, Corchado E (eds) Computational intelligence in security for information systems. Springer, Heidelberg, pp 117–126. https://doi.org/10.1007/978-3-642-04091-7_15
35. Ghiran AM, Silaghi GC, Tomai N (2009) Ontology-based tools for automating integration and validation of firewall rules. In: Abramowicz W (ed) Business information systems. Springer, Heidelberg, pp 37–48. https://doi.org/10.1007/978-3-642-01190-0_4
36. Choraś M, Flizikowski A, Kozik R, Hołubowicz W (2010) Decision aid tool and ontology-based reasoning for critical infrastructure vulnerabilities and threats analysis. In: Rome E, Bloomfield R (eds) Critical information infrastructures security. Springer, Heidelberg, pp 98–110. https://doi.org/10.1007/978-3-642-14379-3_9
37. Miksa K, Sabina P, Kasztelnik M (2010) Combining ontologies with domain specific languages: a case study from network configuration software. In: Aßmann U, Bartho A, Wende C (eds) Reasoning web. Semantic technologies for software engineering. Springer, Heidelberg, pp 99–118. https://doi.org/10.1007/978-3-642-15543-7_4
38. ETSI Industry Specification Group (2013) Measurement ontology for IP traffic (MOI); requirements for IP traffic measurement ontologies development. ETSI, Valbonne. http://www.etsi.org/deliver/etsi_gs/MOI/001_099/003/01.01.01_60/gs_moi003v010101p.pdf
39. Martínez A, Yannuzzi M, Serral-Gracià R, Ramírez W (2014) Ontology-based information extraction from the configuration command line of network routers. In: Prasath R, O'Reilly P, Kathirvalavakumar T (eds) Mining intelligence and knowledge exploration. Springer, Cham, pp 312–322. https://doi.org/10.1007/978-3-319-13817-6_30
40. Martínez A, Yannuzzi M, López J, Serral-Gracià R, Ramírez W (2015) Applying information extraction for abstracting and automating CLI-based configuration of network devices in heterogeneous environments. In: Laalaoui Y, Bouguila N (eds) Artificial intelligence applications in information and communication technologies. Springer, Cham, pp 167–193. https://doi.org/10.1007/978-3-319-19833-0_8
41. Laskey K, Chandekar S, Paris B-P (2015) A probabilistic ontology for large-scale IP geolocation. In: Laskey KB, Emmons I, Costa PCG, Oltramari A (eds) Tenth Conference on Semantic Technology for Intelligence, Defense, and Security. RWTH Aachen University, Aachen, pp 18–25. http://ceur-ws.org/Vol-1523/STIDS_2015_T03_Laskey_etal.pdf
42. Moraes PS, Sampaio LN, Monteiro JAS, Portnoi M (2008) MonONTO: a domain ontology for network monitoring and recommendation for advanced Internet applications users. In: 2008 IEEE Network Operations and Management Symposium Workshops–NOMS 2008. IEEE, Piscataway, NJ, USA. https://doi.org/10.1109/NOMSW.2007.21
43. Sikos LF, Stumptner M, Mayer W, Howard C, Voigt S, Philp D (2018) Representing network knowledge using provenance-aware formalisms for cyber-situational awareness. Procedia Comput Sci 126C: 29–38
44. Sikos LF, Stumptner M, Mayer W, Howard C, Voigt S, Philp D (2018) Automated reasoning over provenance-aware communication network knowledge in support of cyber-situational awareness. In: Liu W, Giunchiglia F, Yang B (eds) Knowledge science, engineering and management. Springer, Cham., pp. 132–143. https://doi.org/10.1007/978-3-319-99247-1_12

Chapter 2
Knowledge Representation of Network Semantics for Reasoning-Powered Cyber-Situational Awareness

Leslie F. Sikos, Dean Philp, Catherine Howard, Shaun Voigt, Markus Stumptner, and Wolfgang Mayer

Abstract For network analysts, understanding how network devices are interconnected and how information flows around the network is crucial to the cyber-situational awareness required for applications such as proactive network security monitoring. Many heterogeneous data sources are useful for these applications, including router configuration files, routing messages, and open datasets. However, these datasets have interoperability issues, which can be overcome by using formal knowledge representation techniques for network semantics. Formal knowledge representation also enables automated reasoning over statements about network concepts, properties, entities, and relationships, thereby enabling knowledge discovery. This chapter describes formal knowledge representation formalisms to capture the semantics of communication network concepts, their properties, and the relationships between them, in addition to metadata such as data provenance. It also describes how the expressivity of these knowledge representation mechanisms can be increased to represent uncertainty and vagueness.

2.1 Introduction

Proactive network security monitoring highly depends on accurate, concise, quality network data. Traditional network vulnerability checking relies on labor-intensive processes that require expertise and are prone to errors.

While hackers' attempts to compromise critical infrastructures are inevitable, intelligent systems provide mechanisms to reduce the impact of cyberattacks, and whenever possible, prevent attacks and manage vulnerability risks via real-time cyber-situational awareness. Transforming raw data into valuable information based

---

L. F. Sikos (✉) · M. Stumptner · W. Mayer
University of South Australia, Adelaide, Australia
e-mail: leslie.sikos@unisa.edu.au

D. Philp · C. Howard · S. Voigt (✉)
Defence Science and Technology Group, Department of Defence,
Australian Government, Adelaide, Australia
e-mail: shaun.voigt@dst.defence.gov.au

© Springer Nature Switzerland AG 2019

L. F. Sikos (ed.), *AI in Cybersecurity*, Intelligent Systems Reference Library 151,
https://doi.org/10.1007/978-3-319-98842-9_2

on the semantics of network devices and the traffic flow between them cannot rely solely on manual methods because of the data volume and diversity that characterize communication networks. Network analysts can benefit from automated reasoning-based frameworks to gain improved cyber-situational awareness of the highly dynamic cyberspace [1]. Developing such frameworks requires data integration via syntactic and semantic interoperability, which can be achieved by formal knowledge representation standards, such as the *Resource Description Framework (RDF)*[1] [2].

This chapter describes formal knowledge representation formalisms to capture the semantics of communication network concepts, their properties, and the relationships between them. After introducing some basic communication network concepts (Sect. 2.3), a detailed discussion explains how to represent these (Sect. 2.4), along with metadata (Sect. 2.5), including data provenance, and how to increase expressivity to be able to represent uncertainty and vagueness (Sects. 2.6 and 2.7).

2.2 Preliminaries

One technique to enable automated data processing for communication networks is to provide syntactic and semantic interoperability, which can be achieved through *formal knowledge representation*. Formal knowledge representation is a field of artificial intelligence (AI), which is dedicated to representing information about a selected part of the world, called the *knowledge domain* (field of interest or area of concern), in a form that allows software agents to solve complex tasks. Formally represented information can not only be indexed and queried very efficiently, but also enable reasoning algorithms to infer new statements, thereby facilitating *knowledge discovery*. From the representation perspective, there are two types of network data: expert knowledge and statements about entities (individuals) of real-world networks.

Expert knowledge represents networking concepts and relationships that can be codified as terminological definitions in *ontologies* as background knowledge, thereby describing ground truth about communication networks, such as IP addresses should not be reused within a public network. Ontology-based representations use a common data format that enables the information fusion needed to maximize semantic enrichment [3]. Owing to the complexity of communication networks, there is a variety of network ontologies, all of which capture the semantics of different aspects of networking. Some notable network ontologies include the *Physical Devices Ontology* [4], the *Network Domain Ontology* [5], the *Ontology for Network Device Configuration (ONDC)* [6], the *Common Information Model Ontology (CIM)* [7], the *Ontology for the Router Configuration Domain (ORCONF)* [8], a probabilistic ontology for large-scale IP geolocation [9], the *Measurement Ontology for IP Traffic (MOI)* [10], and the *Network Ontology (NetOnto)* [11]. However, not all network ontologies are suitable for representing network topology and network traffic

[1] https://www.w3.org/RDF/

flow because of their scope, conceptualization, and intended application [12]. This led to the development of the *Communication Network Topology and Forwarding Ontology (CNTFO)*[2] [13].

Entities of real-world networks, which can be used to describe a particular network, can be characterized by properties that are either declared manually or extracted by software agents from routing messages[3] sent between network devices, configuration files of network devices, etc. Many data sources utilized for cyber-situational awareness, such as traceroute responses, routing messages, and router configuration files, constitute *unstructured data*, which is often stored in proprietary formats. The corresponding information is human-readable only, and only experts can comprehend it. For example, software agents cannot efficiently process the natural sentence the "IP address of ServerX is 103.254.136.21," because to software agents this statement would be just a meaningless string. In contrast, the same statement written in a *semistructured data* format, such as XML,[4] is machine-readable, because the name of the device and the IP address are annotated with separate tags (e.g., `<device>ServerX</device>` and `<ipadd>103.254.136.21</ipadd>`), which allows software agents to extract the corresponding data. However, the *semantics* (meaning) of these properties are still not defined. To overcome this limitation, the statement can be written as *structured data* using *Semantic Web standards*, such as RDF, and extended with annotations from knowledge organization systems (KOS), in particular *controlled vocabularies*, which collect terms, properties, relationships, and entities of a knowledge domain, and *ontologies*, which are formal conceptualizations of a knowledge domain with complex classes, relationships, and rules suitable for inferring new statements. By doing so, the data becomes machine-interpretable, unambigious, and interoperable, allowing a much wider range of automated tasks than unstructured or semistructured data. Structured data written according to best practices for publishing structured data is called *Linked Data*. Linked Data is often used to complement concept definitions of ontologies to further detail the semantics of a knowledge representation and interlink related resources [14]. The Linked Data principles include dereferencable URIs for every resource over the HTTP protocol, knowledge representation using open standards, and links to related resources [15]. If published with an open license (e.g., PDDL,[5] ODC-By,[6] ODC-ODbL,[7] CC0 1.0,[8] CC-BY-SA,[9] GFDL),[10] Linked Data becomes *Linked Open Data (LOD)*. Those datasets that contain Linked Open Data are called *LOD datasets*.

---

[2]http://purl.org/ontology/network/

[3]*Routing* is the process of selecting network paths to carry network traffic.

[4]https://www.w3.org/TR/xml11/

[5]https://opendatacommons.org/licenses/pddl/

[6]https://opendatacommons.org/licenses/by/

[7]https://opendatacommons.org/licenses/odbl/

[8]https://creativecommons.org/publicdomain/zero/1.0/

[9]https://creativecommons.org/licenses/by-sa/4.0/

[10]https://www.gnu.org/copyleft/fdl.html

RDF allows machine-interpretable statements of the form subject-predicate-object (see Definition 2.1).

Definition 2.1 (**RDF Triple**) Let S be a set of data sources, which is a subset of the set of International Resource Identifiers (IRIs) ($\mathbb{I}$), i.e., sets of strings of Unicode characters of the form scheme:[//[user:password@]host[:port][/]path [?query][#fragment], formally $S \subset \mathbb{I}$. Assume there are pairwise disjoint infinite sets of

1. IRIs;
2. RDF literals ($\mathbb{L}$), which are either
   a. self-denoting plain literals $\mathbb{L}_P$ of the form "<string>"(@<lang>), where <string> is a string and <lang> is an optional language tag; or
   b. typed literals $\mathbb{L}_T$ of the form "<string>"^^<datatype>, where <datatype> is an IRI denoting a datatype according to a schema, such as the XML Schema, and <string> is an element of the lexical space corresponding to the datatype; and
3. blank nodes ($\mathbb{B}$), i.e., unique but anonymous resources that are neither IRIs nor RDF literals.

A triple $(s, p, o) \in (\mathbb{I} \cup \mathbb{B}) \times \mathbb{I} \times (\mathbb{I} \cup \mathbb{L} \cup \mathbb{B})$ is called an *RDF triple* (or RDF statement), where s is the subject, p is the predicate, and o is the object.

For example, the fact that routers are networking devices can be expressed as shown in Example 2.1.

Example 2.1 A triple

Router–isA–NetworkingDevice

To declare the definition of each triple element, use the definition of "Router" from DBpedia, the "isA" relationship from the RDF vocabulary (rdfV) and the definition of networking devices from DBpedia. This leads to the following RDF triple:

- Subject: http://dbpedia.org/resource/Router_(computing)
- Predicate: http://www.w3.org/1999/02/22-rdf-syntax-ns#type
- Object: http://dbpedia.org/resource/Networking_ device

Long and frequently used URLs can be abbreviated using the namespace mechanism, which defines a prefix URL that can be concatenated with the triple element to obtain the full URL. Because RDF supports a wide range of serialization formats, RDF triples can be represented in a variety of ways. Using the Turtle serialization,[11] for example, the previous triple would look like Listing 2.1.

[11] https://www.w3.org/TR/turtle/

Listing 2.1 An RDF triple

```
@prefix dbpedia: <http://dbpedia.org/resource/> .
@prefix rdf: <http://www.w3.org/1999/02/22-rdf-syntax-ns#> .

dbpedia:Router_(computing) rdf:type dbpedia:Networking_device .
```

A set of RDF triples can be visualized as a directed, labeled graph, called an *RDF graph*, in which the set of nodes corresponds to the set of subjects and objects of RDF triples in the graph, and the edges represent the predicate relationships that hold between the subjects and the objects.

RDF triples complemented by context form RDF quadruples (see Definition 2.2).

Definition 2.2 (**RDF Quadruple**) A 4-tuple of the form (s, p, o, c), where s, p, o represents an RDF triple and c identifies the context of the triple, is called an *RDF quadruple* (RDF quad).

For example, `:host18 :hasInterface :I10.1.8.1 :traceroute101.` is an RDF quad, which describes that host18 has the interface address 10.1.8.1, according to traceroute #101.

The context of each triple can identify the RDF graph to which the triple belongs (see Definition 2.3).

Definition 2.3 (**Named Graph**) If the context is an IRI that identifies the name of the RDF graph the triple belongs to, an RDF dataset D can be defined as a set of RDF graphs of the form $g_0, n_1, g_1, \ldots, n_m, g_m$, where each g_i is a graph and g_0 is the default graph of D. Each optional pair n_i, g_i is called a *named graph*, in which $n_1, \ldots, n_m \in \mathbb{I}$ are unique graph names in D [16].

To achieve a favorable trade-off between expressivity and reasoning complexity, and to ensure decidability,[12] knowledge representations can be formally grounded in *description logics* (DL) [17]. DL formalisms are typically implemented in RDF statements that utilize terms from ontologies written in the Web Ontology Language (OWL).[13] The expressivity of each description logic is determined by the supported mathematical constructors, which is reflected by the letters in the DL names. For example, the core description logic $\mathcal{ALC}$ (Attributive Language with Complements) allows atomic and complex concept negation, concept intersection, universal restrictions, and limited existential quantification. The extension of $\mathcal{ALC}$ with transitive roles is $\mathcal{S}$. $\mathcal{S}$ extended with role hierarchies ($\mathcal{H}$), inverse roles ($\mathcal{I}$), functional properties ($\mathcal{F}$), and datatypes ($\mathcal{D}$) is called $\mathcal{SHIF}^{(\mathcal{D})}$, which roughly corresponds to OWL Lite. Adding nominals ($\mathcal{O}$) and cardinality restrictions ($\mathcal{N}$) to $\mathcal{SHIF}^{(\mathcal{D})}$ leads to $\mathcal{SHOIN}^{(\mathcal{D})}$, the description logic behind OWL DL. $\mathcal{SROIQ}^{(\mathcal{D})}$ is the logical

[12]The decidability of a formalism ensures that an inference algorithm will not run into an infinite loop.

[13]https://www.w3.org/OWL/

underpinning of OWL 2 DL, which, beyond the $\mathcal{SHOIN}^{(\mathcal{D})}$ constructors, allows the union of a finite set of complex role inclusions and a role hierarchy ($\mathcal{R}$), and replaces the unqualified cardinality restrictions ($\mathcal{N}$) with qualified cardinality restrictions ($\mathcal{Q}$).

OWL 2 DL-compliant knowledge bases, with a logical underpinning in the $\mathcal{SROIQ}^{(\mathcal{D})}$ description logic, are based on three finite and pairwise disjoint sets of atomic concepts (N_C), atomic roles (N_R), and individual names (N_I), where (N_C), (N_R), and (N_I) $\in \mathbb{I}$. They may implement a wide range of constructors in concept expressions, including concept names, concept intersection, concept union, complement, tautology, contradiction, existential and universal quantifiers, qualified at-least and at-most restrictions, local reflexivity, and individual names (see Definition 2.4).

Definition 2.4 **($\mathcal{SROIQ}$ Concept Expression)** The set of constructors allowed in $\mathcal{SROIQ}$ concept expressions can be defined as $\mathbf{C} ::= N_C \mid (C \sqcap D) \mid (C \sqcup D) \mid \neg C \mid \top \mid \bot \mid \exists R.C \mid \forall R.C \mid \geqslant nR.C \mid \leqslant nR.C \mid \exists R.Self \mid N_I$, where C represents concepts, R represents roles, and n is a nonnegative integer.

The role expressions support the universal role, atomic roles, and negated atomic roles (see Definition 2.5).

Definition 2.5 **($\mathcal{SROIQ}$ Role Expression)** The permissible constructors of $\mathcal{SROIQ}$ role expressions can be defined as $\mathbf{R} ::= U \mid N_R \mid N_{R^-}$.

Based on these sets, each axiom is either a concept inclusion, an individual assertion, or a role assertion (see Definition 2.6).

Definition 2.6 **($\mathcal{SROIQ}$ Axiom)** $\mathcal{SROIQ}$ axioms can be classified as follows:

- general concept inclusions of the form $C \sqsubseteq D$ and $C \equiv D$ for concepts C and D (terminological knowledge, TBox);
- individual assertions of the form $C(a)$, $R(a, b)$, $\neg R(a, b)$, $a \approx b$, $a \not\approx b$, where a, $b \in N_I$ denotes individual names, $C \in \mathbf{C}$ denotes concept expressions, and $R \in \mathbf{R}$ denotes roles (assertional knowledge, ABox);
- role assertions of the form $R \sqsubseteq S$, $R_1 \circ \ldots \circ n \sqsubseteq S$, Asymmetric($R$), Reflexive($R$), Irreflexive($R$), or Disjoint($R$, S) for roles R, R_i, and S (role box, RBox).

The meaning of formally described network concepts and roles is defined by their model-theoretic semantics, which are based on interpretations. These interpretations consist of a domain of discourse, Δ, consisting of two disjoint sets and an interpretation function. An object domain $\Delta^{\mathcal{I}}$ covers specific abstract objects (individuals) and classes of abstract objects (concepts) as well as abstract roles. A concrete domain $\mathcal{D}$ is a pair ($\Delta_{\mathcal{D}}$, $\Phi_{\mathcal{D}}$), where $\Delta_{\mathcal{D}}$ is a set called the datatype domain, and $\Phi_{\mathcal{D}}$ is a set of concrete predicates. Each predicate name P from $\Phi_{\mathcal{D}}$ is associated with an arity n, and an n-ary predicate $P^D \subseteq \Delta_{\mathcal{D}}^n$. Concrete domains integrate description logics with concrete sets, such as natural numbers ($\mathbb{N}$), integers ($\mathbb{Z}$), real numbers ($\mathbb{R}$), complex numbers ($\mathbb{C}$), strings, Boolean values, and date and time values, along with concrete roles defined on these sets, including numerical comparisons, string comparisons, and comparisons with constants. A Tarski-style DL interpretation $\mathcal{I}$ = $(\Delta^{\mathcal{I}}, \cdot^{\mathcal{I}})$ utilizes a concept interpretation function $\cdot^{\mathcal{I}_C}$ to map concepts into subsets of the object domain, a role interpretation function $\cdot^{\mathcal{I}_R}$ to map object roles into subsets of $\Delta^{\mathcal{I}} \times \Delta^{\mathcal{I}}$ and datatype roles into subsets of $\Delta^{\mathcal{I}} \times \Delta_{\mathcal{D}}$, and an individual interpretation function $\cdot^{\mathcal{I}_I}$ to map individuals into elements of the object domain.

2.3 Communication Network Concepts

The following sections briefly introduce some communication networking concepts and discuss how the semantics of these concepts and their properties and relationships can be captured using formal knowledge representation formalism. While core networking concepts can be described using expert knowledge, as evidenced by a large number of networking ontologies, the definition of networking properties is far less trivial. Firstly, not all properties are important for understanding how network devices are interconnected or how information flows around the network; the relevant properties must be carefully selected. Secondly, not all networking properties are standardized, and even the ones that are may have proprietary implementations, as seen in the implementations of routing protocols in Cisco and Juniper routers. Thirdly, the definition of complex network properties requires custom datatypes, not all of which can be defined by restricting standardized datatypes, and some of which are context-dependent. The following sections discuss some of the interdependencies and properties of core networking concepts from the formal representation point of view.

2.3.1 Networks and Topologies

Communication networks establish connections between network nodes to exchange data via cable or wireless media. These network nodes are either hosts, such as personal computers, workstations, and servers, or network devices, such as modems, routers, switches, repeaters, hubs, bridges, gateways, VPN appliances, and firewalls. In knowledge representation, these network nodes can be described by concept definitions, which can be used in arbitrary statements. In a network, if multiple instances of the same node type are present, individual names typically contain consecutive numbers after the node type, such as `Router1` and `Router2`. If the name of the actual network device is known, such as the name used in routing messages, the individual name will be identical to that name, e.g., `Router10143R1`.

The *topology* of a communication network is the arrangement of the various network elements, such as routers, computers, and links, within the network. There are two basic categories of computer network topologies, *physical topologies* and *logical topologies*. The physical topology of a network is the arrangement of the physical components of the network, including the location of devices and cables, while the logical network topologies show how network devices logically communicate with each other via a communication protocol. Network topologies can be described formally by the nodes and the relationships between them, as for example, by using the `connectedTo` predicate between directly connected devices. Each node in a representation must have a unique identifier, such as a public IP address or a name consisting of the node type and a number. Sophisticated statements about complex networks typically need a combination of background knowledge and individual assertions.

2.3.2 Network Interfaces and IP Addressing

The *Internet Protocol (IP)* is the fundamental network routing protocol used in communication networks. To participate in an IP network, each network element must have a properly configured interface to the IP network. A properly configured interface has two main parameters: an IP address and a subnet mask. IP addresses are 32 bits long, and are unique within the whole IP network. To make IP addresses and subnet masks easy to read, they are often written in dotted decimal format [26] (see Fig. 2.1).

The subnet mask defines the Network ID and Host ID portions of an IP address. Property-value pairs can be used in formal representations to describe network interfaces. Network interfaces from different network elements are joined together into sub-networks (subnets). Different subnets are joined together to form a larger IP network. A network element that joins one or more subnets together is called a router.

2.3.3 Routers

When a router receives a packet that is not destined for one of its own interfaces, it uses a routing table to determine to which interface to forward that packet towards its destination. For each reachable destination, a routing table lists which connected

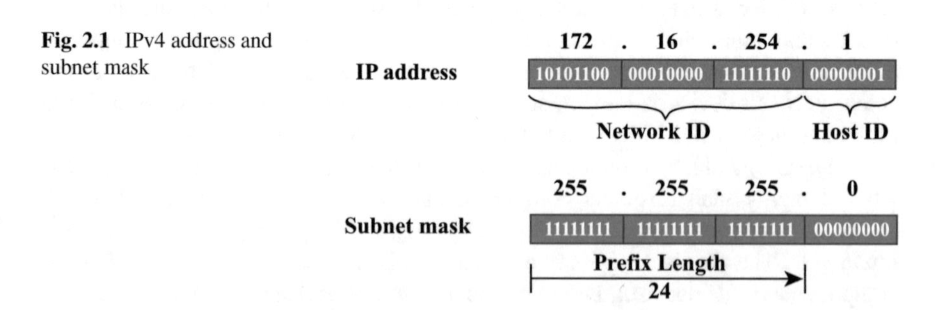

Fig. 2.1 IPv4 address and subnet mask

network element is next along the path to the destination. When an IP packet arrives, a router uses this table to determine the interface[14] on which to forward the packet based on its destination IP address.[15] The next routers then repeat this process using their own routing table until the packet reaches its destination. For an IP network to function, the routing tables on all of the network elements need to be configured in a consistent manner with routing table entries that forward packets from any network element to any other network element.

Each router is configured using a router configuration file. These files hold all the commands and parameters required to configure the router [18]. To demonstrate the types of data that can be used from router configurations to gain cyber-situational awareness, consider the fragment of a Cisco router configuration file shown in Listing 2.2.

Listing 2.2 Part of a router configuration file [19]

```
no keepalive
no cable proxy-arp
cable helper-address 10.100.0.30
interface FastEthernet0/0
ip address 10.100.0.14 255.255.255.0
no ip directed-broadcast
no ip mroute-cache
!!
interface Ethernet2/3
ip address 10.145.30.22 255.255.255.0
no ip directed-broadcast
!
interface Cable5/0
ip address 172.1.71.1 255.255.255.0 secondary
ip address 10.100.1.1 255.255.252.0
no ip directed-broadcast
no ip route-cache
no ip mroute-cache
cable downstream annex B
cable downstream modulation 64qam
cable downstream interleave-depth 32
cable upstream 0 spectrum-group 1
cable upstream 0 modulation-profile 3
cable downstream frequency 531000000
cable upstream 0 frequency 28000000
cable upstream 0 power-level 0
no cable upstream 0 shutdown
cable upstream 1 shutdown
cable upstream 2 shutdown
cable upstream 3 shutdown
```

[14]There is usually one-to-one mapping between interfaces and IP addresses.

[15]Within a subnet, each IP address is assumed to be unique.

2.3.4 Autonomous Systems and Routing

An *autonomous system (AS)* is a network, or collection of networks, managed and supervised by a single administrative entity or organization. Such an administrative entity can be an Internet service provider, an enterprise, a university, a company division, or a group of organizations. For the purpose of routing, each AS is assigned a globally unique AS number, called the *Autonomous System Number (ASN)*, assigned by the Regional Internet Registry (RIR) of the country of registration (e.g., APNIC[16] in the Asia Pacific). For example, ASN 3356 identifies Level 3 Communications, Inc., ASN 1299 corresponds to Telia Company AB, and ASN 174 represents Cogent Communications. ASNs can be used in individual names to make statements about ASes and their relationships with organizations, network devices, and network events.

Different protocols are used to route traffic between and within ASes. The routing of traffic between ASes (which is referred to as *inter-AS routing*) is managed by Exterior Gateways Protocols (EGPs), such as BGP [20]. BGP is used to facilitate inter-AS relationships by exchanging routing and reachability information among ASes on the Internet. BGP is commonly classified as a path vector routing protocol. *Autonomous system boundary routers* (ASBRs) advertise network reachability information (that is, the networks which are reachable via each of their neighbors and how many hops away each network is). When a BGP session is initialized between routers, messages are sent to exchange routing information until the complete BGP routing table has been exchanged. BGP makes routing decisions based on AS paths, network policies, and rulesets configured by the network administrators. A *BGP peer group* is a set of BGP neighbors that shares the same outbound routing policy (although the inbound policies may differ).

The routing of traffic within an individual AS (which is referred to as *intra-AS routing*) is managed by an Interior Gateway Protocol (IGP), such as the OSPF protocol [21], the Intermediate System to Intermediate System (IS-IS) protocol [22], the Routing Information Protocol (RIP) [23] and the Enhanced Interior Gateway Routing Protocol (EIGRP). Within an AS, multiple IGPs can be used simultaneously and multiple instances of the same IGP can be used on different network segments. Routing tables can be configured manually or populated using dynamic routing protocols such as BGP, OSPF, IS-IS, and RIP.

OSPF, which is a link state routing protocol,[17] is the most widely used interior gateway protocol on the Internet [24]. Each link has an associated cost metric, where the cost is often determined by factors such as the speed of the router's interface, data throughput, link availability, and link reliability. Each participating router communicates its own link states to every other participating router such that each router has an identical view of the entire OSPF network. Using the topology constructed from the advertised link state information, and Dijkstra's *shortest path first algorithm* [25], each router computes the shortest path tree for each destination network.

---

[16]https://www.apnic.net

[17]In a link state routing protocol, each router constructs a map of the connectivity of the network in which it resides.

This tree is then used to create the routing table, which is then used to determine the interface on which to forward the individual incoming packets. Because link state routing protocols, like OSPF, maintain information about the complete topology, this information can be exploited to support cyber-situational awareness [27].

Each OSPF router within a network is assigned a Router Identification Number. This number, which is hereafter referred to as *routerID*, uniquely identifies the router within a routing domain.[18] The routerID is displayed using the dot-decimal notation, for example, 10.0.2.2. OSPF networks are often subdivided into *OSPF areas* to simplify administration or optimize traffic flow or resource utilization. These routing areas are identified by 32-bit numbers, which is expressed either in decimal or the octet-based dot-decimal notation typical to IPv4 addresses. The core or backbone area of an OSPF network is Area 0 or 0.0.0.0.

Routers can be classified according to their location and role in the network. The OSPF protocol itself is configured on a router's interface, rather than the router as a whole. Those routers that have interfaces in multiple areas are called *area border routers* (ABRs). Routers that have an interface in the backbone area are called *backbone routers*. Routers that have interfaces in only one area are called *internal routers*. Routers that exchange routing information with routing devices in non-OSPF networks are called *autonomous system boundary routers* (ASBRs). ASBRs advertise externally learned routes throughout the OSPF AS. Depending on the location of an ASBR in the network, it can be an ABR, a backbone router, or an internal router. These different router types can be defined as classes and used with the `rdf:type` predicate to make statements about the router type. *Link State Advertisements (LSAs)* are the basic communication mechanism of the OSPF routing protocol. There are eleven different types of LSAs. Some of the information included in these LSAs is useful for cyber-situational awareness.

2.4 Formal Knowledge Representation for Cyber-Situational Awareness

The following sections demonstrate the formal representation of the communication network concepts introduced in the previous section.

[18]In computer networking, a routing domain is a collection of networked systems that operate common routing protocols and are under the control of a single administrative entity. A given AS may contain multiple routing domains. A routing domain can exist without being an Internet-participating AS.

2.4.1 Representing Network Knowledge Using Ontology Definitions

To demonstrate how core networking concepts and properties can be defined formally in ontologies, take a look at the description logic definition of the router configuration properties shown in Listing 2.3.

Listing 2.3 Formal definition of some router configuration terms with description logic axioms

Area
Interface
Network
NetworkElement
OSPFSummaryRouteEntry
Router
$\mathrm{Router} \sqsubseteq \mathrm{NetworkElement}$
RouteEntry
connectedTo
$\exists \mathrm{connectedTo}.\top \sqsubseteq \mathrm{Interface}$
$\top \sqsubseteq \forall \mathrm{connectedTo}.\mathrm{Network}$
$\top \sqsubseteq \leqslant 1 \mathrm{areaId}.\top$
$\exists \mathrm{areaId}.\top \sqsubseteq \mathrm{Area}$
$\top \sqsubseteq \forall \mathrm{areaId}.\mathrm{unsignedInt}$
$\exists \mathrm{ip}.\top \sqsubseteq \mathrm{Interface}$
$\mathrm{ipv4} \sqsubseteq \mathrm{ip}$
Dis(ipv4, areaId)
Dis(ipv4, routerId)
$\exists \mathrm{ipv4}.\top \sqsubseteq \mathrm{Interface}$
$\top \sqsubseteq \forall \mathrm{ipv4}.\mathrm{ipv4Type}$
$\exists \mathrm{routerId}.\top \sqsubseteq \mathrm{RouteEntry} \sqcup \mathrm{Router} \sqcup \mathrm{OSPFSummaryRouteEntry}$
$\top \sqsubseteq \forall \mathrm{routerId}.\mathrm{ipv4Type}$
ipv4Type ≡ string["(([0−1]?[0−9]?[0−9]|2([0−4][0−9]|5[0−5]))
\.){3}([0−1]?[0−9]?[0−9]|2([0−4][0−9]|5[0−5]))"]

These definitions have been implemented in CNTFO as shown in Listing 2.4.

Listing 2.4 Some router configuration concepts and properties defined in RDF (Turtle syntax)

```
@prefix net: <http://purl.org/ontology/network/> .
@prefix rdf: <http://www.w3.org/1999/02/22-rdf-syntax-ns#> .
@prefix rdfs: <http://www.w3.org/2000/01/rdf-schema#> .
@prefix owl: <http://www.w3.org/2002/07/owl#> .
@prefix xsd: <http://www.w3.org/2001/XMLSchema#> .

:Area a owl:Class .
:Interface a owl:Class .
:Network a owl:Class .
:NetworkElement a owl:Class .
:OSPFSummaryRouteEntry a owl:Class .
:Router a owl:Class ; rdfs:subclassOf :NetworkElement .
:RouteEntry a owl:Class .

:connectedTo a owl:ObjectProperty ; rdfs:domain :Interface ;
    rdfs:range :Network .
```

```
:areaId a owl:DatatypeProperty , owl:FunctionalProperty ;
rdfs:domain :Area ; rdfs:range xsd:unsignedInt .
:ip a owl:DatatypeProperty ; rdfs:domain :Interface .
:ipv4 a owl:DatatypeProperty ; rdfs:subPropertyOf :ip ;
owl:propertyDisjointWith :areaId , :routerId ; rdfs:domain
:Interface ; rdfs:range :ipv4Type .
:routerId a owl:DatatypeProperty ; rdfs:domain [ a owl:Class ;
    owl:unionOf (:RouteEntry :Router :OSPFSummaryRouteEntry) ]
    ; rdfs:range net:ipv4Type .

:ipv4Type a rdfs:Datatype ; owl:onDatatype xsd:string ;
owl:withRestrictions ( [xsd:pattern
    "(([0-1]?[0-9]?[0-9]|2([0-4][0-9]|5[0-5]))\.){3}
    ([0-1]?[0-9]?[0-9]|2([0-4][0-9]|5[0-5]))"] ) .
```

These definitions can be used in arbitrary RDF statements, in which the domain, range, and datatype restrictions apply accordingly. Assume the fragment of a router configuration file shown in Listing 2.5.

Listing 2.5 Some properties with their values in a router configuration file

```
#---- Router configuration file for RE20
hostname "RE20"
interface FastEthernet0/0 ip address 10.100.0.13 255.255.255.252
router ospf 1 router-id 10.100.0.13

#---- Router configuration file for RE37
hostname "RE37"
interface FastEthernet0/0 ip address 10.100.0.14 255.255.255.252
router ospf 1 router-id 10.100.0.14
```

These can be written in a description logic formalism as shown in Listing 2.6.

Listing 2.6 Formal description of Listing 2.5 with description logic axioms

Router(RE20)
RE20 ⊓ ∃hostname.{RE20}
RE20 ⊓ ∃routerId.{10.100.0.13}
Interface(RE20_FastEthernet0_0)
RE20_FastEthernet0_0 ⊓ ∃ipv4.{10.100.0.13}
Router(RE37)
RE20 ⊓ ∃hostname.{RE37}
RE37 ⊓ ∃routerId.{10.100.0.14}
Interface(RE37_FastEthernet0_0)
RE37_FastEthernet0_0 ⊓ ∃ipv4.{10.100.0.14}
Network(N10.100.0.12/30)
N10.100.0.12/30 ⊓ ∃ipv4Subnet.{10.100.0.12/30}

This can be serialized in Turtle using the concept, role, and datatype definitions from CNTFO as shown in Listing 2.7.

Listing 2.7 Turtle serialization of Listing 2.6

```
# RE20 is a router and defines
# hostname, interface, ipv4, and OSPF routerId
:RE20 a net:Router .
:RE20 net:hostname "RE20" .
:I10.100.0.13 a net:Interface .
:I10.100.0.13 net:ipv4 "10.100.0.13"^^net:ipv4Type .
:RE20 net:hasInterface :I10.100.0.13 .
:RE20 net:routerId "10.100.0.13"^^net:ipv4Type .

# RE37 is a router and defines
# hostname, interface, ipv4, and OSPF routerId
:RE37 a net:Router .
:RE37 net:hostname "RE37" .
:I10.100.0.14 a net:Interface .
:I10.100.0.14 net:ipv4 "10.100.0.14"^^net:ipv4Type .
:RE37 net:hasInterface :I10.100.0.14 .
:RE37 net:routerId "10.100.0.14"^^net:ipv4Type .

# ipv4 addresses belong to subnets
# subnets define networks; 255.255.255.252 is a /30 network
# Both RE20 and RE27 interfaces have ipv4 addresses
:N10.100.0.12_30 a net:Network .
:N10.100.0.12_30 net:ipv4subnet
"10.100.0.12/30"^^net:ipv4SubnetType .
:I10.100.0.13 net:connectedTo :N10.100.0.12_30 .
:I10.100.0.14 net:connectedTo :N10.100.0.12_30 .
```

This example shows how we can define the network connection between two routers, which corresponds to the red link in Fig. 2.2.

We define the two routers, `:RE20` and `:RE37`, with hostnames `RE20` and `RE37` using `net:hostname`. Interfaces `:I10.100.0.13` (for RE20) and `:I10.100.0.14` (for RE37) are attached via the `net:hasInterface` predicate. Network interfaces have IPv4 addresses, defined by the `net:ipv4` datatype property of an interface. Because IPv4 addresses are defined in dotted notation format, we restrict the strings "10.100.0.13" and "10.100.0.14" to `net:ipv4Type`. Since IPv4 networks are defined in terms of IPv4 subnets, we also define the network `:N10.100.0.12_30`, with an IPv4 address range defined by `net:ipv4subnet`. The `net:ipv4SubnetType` restriction ensures that the string "10.100.0.12/30" is valid. To complete our network connection between RE20 and RE37, we state that their respective interfaces are connected to the same subnet. Specifically, both `:I10.100.0.13` and `:I10.100.0.14` are `net:connectedTo` `:N10.100.0.12_30`.

In summary, using Turtle serialization, we have defined the pattern Router-Interface-Network-Interface-Router. This pattern of network connectivity between routers was derived from the raw file in Listing 2.5, then with formal semantics in Listing 2.6, then in Turtle format in Listing 2.7. This way, we can formally describe the semantics of the full network of Fig. 2.2 using Turtle and CNTFO. Moreover, because we express network connectivity formally using CNTFO, we can define rules

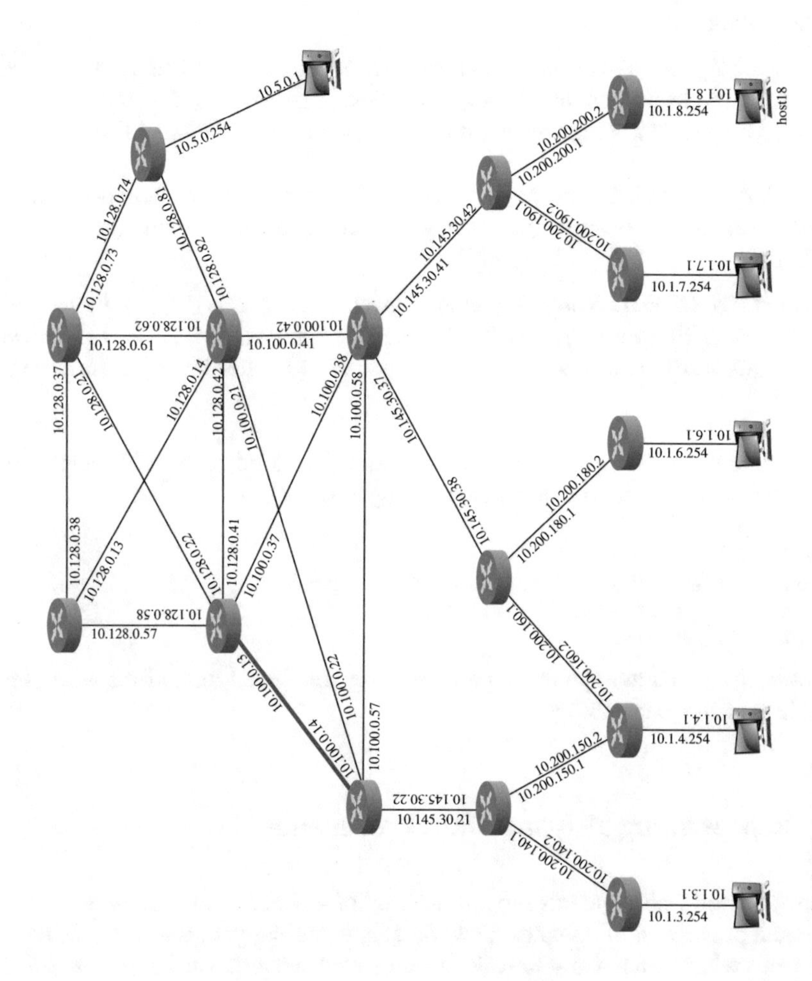

Fig. 2.2 Topology of an enterprise-level network

that facilitate information fusion and enable automated reasoning across multiple disparate network data sources. This automation is largely hidden from the network analysts, while the information it provides significantly increases cyber-situational awareness.

Complex routing policies may need an expressivity higher than that of OWL, such as SWRL[19] rules. SWRL rules use unary predicates for describing classes and data types, binary predicates for properties, and built-in n-ary predicates (see Definition 2.7).

Definition 2.7 (Atom) An *atom* is an expression of the form $P(\text{arg}_1, \text{arg}_2, \ldots)$, where P is a predicate symbol (classes, properties, or individuals) and $\text{arg}_1, \text{arg}_2, \ldots$ are the arguments of the expression (individuals, data values, or variables).

SWRL rules contain an antecedent (body) and a consequent (head), both of which can be an atom or a positive (i.e., unnegated) conjunction of atoms (see Definition 2.8).

Definition 2.8 (SWRL Rule) A *rule* R is given as $B_1 \wedge \ldots \wedge B_m \rightarrow H_1 \wedge \ldots \wedge H_n (n \geqslant 0, m \geqslant 0)$, where $B_1, \ldots, B_m, H_1, \ldots H_n$ are atoms, $B_1 \wedge \ldots \wedge B_m$ is called the body (premise or antecedent), and $H_1 \wedge \ldots \wedge H_n$ is the head (conclusion or consequent).

For example, a BGP peer group can be defined as a group of BGP neighbors sharing the same outbound policies (see Listing 2.8) [28].

Listing 2.8 A SWRL rule

```
ComputerSystem (?CS) ∧ Dedicated (?CS, 'router')
∧ RouterInAS (?CS,#AS400) → InBGPPeerGroup
(?CS,#AS100_AS400)
```

However, DL axioms should be preferred over rule-based formalisms whenever possible to ensure decidability.

2.5 Representing Network Data Provenance

While Semantic Web standards support the level of task automation required for the automated processing of complex network data, capturing the provenance of RDF statements, which can make automatically generated network data authoritative and verifiable, is not trivial. You see, most mechanisms for capturing provenance in RDF have multiple issues. *Reification*, which makes statements about RDF statements, has no formal semantics, uses blank nodes that cannot be identified globally, and leads to triple bloat, application-dependent interpretation, and difficult-to-query representations. *N-ary relations* also employ blank nodes to represent instances of relations,

[19]Semantic Web Rule Language.

and the restrictions that define the corresponding classes require a large number of triples. Therefore, alternatives have been proposed over the years (some for capturing arbitrary metadata, not provenance specifically), all with different representation prerequisites, provenance granularity and precision, and reasoning complexity. Approaches that represent provenance with triples by extending the standard RDF data model include *RDF/XML Source Declaration*,[20] *Annotated RDF (aRDF)* [29], *Provenance Context Entity* (PaCE) [30], *singleton property* [31], and *RDF** [32]. At the description logic level, an approach called *contextual annotation* [33] was proposed. Quadruple-based approaches include *N-Quads*,[21] *named graphs* [34] and their derivations, such as *RDF triple coloring* [35], *RDF/S graphsets* [36], *nanopublications* [37], as well as *Annotated RDF Schema* [38]. *RDF*$^+$ is an approach that employs quintuples to capture provenance [39]. At the implementation level, many graph databases have proprietary solutions, e.g., AllegroGraph provides sorted index groups and triple attributes that can be used for capturing data provenance [40].

Because named graphs constitute the most standard-compliant means of capturing RDF data provenance, the following examples will demonstrate the use of named graphs. By using two named graphs, the default (or assertion) graph can describe the RDF statements about network knowledge, and the provenance graph can detail the provenance of the network knowledge as shown in Listing 2.9.

Listing 2.9 Provenance-aware network knowledge representation using two named graphs

```
# Network knowledge graph @ http://example.com/NetworkDataset1

@prefix : <http://example.com/NetworkDataset1/> .
@prefix net: <http://purl.org/ontology/network/> .
:Router10143R1 net:inAS :AS10143 .

# Provenance graph @ http://example.org/provenance/

@prefix prov: <http://www.w3.org/ns/prov#> .
<http://example.com/NetworkDataset1/> prov:wasDerivedFrom
<http://example.com/NtwDiscovery20180626/> .
```

The next example, shown in Listing 2.10, utilizes multiple named graphs to capture the provenance of a BGP update message, an OSPF LSA, a router configuration file, and an open dataset used for generating the statements about a router and the AS to which it belongs.

Listing 2.10 Multiple named graphs for representing network knowledge derived from diverse resources

```
@prefix : <http://www.example.org/networkdataset24#> .
@prefix net: <http://purl.org/ontology/network/> .
@prefix prov: <http://www.w3.org/ns/prov#> .

:BGPUM { :Router10143R1 net:inAS :AS10143 . }
:OSPFLSA { :Router10143R1 net:isASBR "true" . }
```

[20] https://www.w3.org/Submission/rdfsource/

[21] https://www.w3.org/TR/n-quads/

```
:R1CF { :Router10143R1 net:hostname "AS10143R1" . }
:OD { :AS10143 net:asName "EXETEL-AS-AP" . }

:PROV { :BGPUM prov:wasDerivedFrom :BGP10143R1toAS1221R1 ;
          :OSPFLSA prov:wasDerivedFrom :OSPF143R1143R2 ;
          :R1CF prov:wasDerivedFrom :AS10143R1ConfigFile ;
          :OD prov:wasDerivedFrom :CAIDAASDataset . }

:META { :PROV prov:generatedAtTime
          "2017-11-02T12:37:00+09:30"^^xsd:dateTime . }
```

Because the various parsers utilized to generate the network data may run at different intervals, the timestamps of the corresponding RDF triples vary. To be able to query massive network datasets efficiently by creation date and time, the provenance for each node have to be distinguished by a unique identifier, which can be used to capture the second layer of provenance, thereby segregating the timestamps from the network data and the rest of the provenance data (see Listing 2.11).

Listing 2.11 Capturing dual-layer provenance with named graphs

```
@prefix : <http://www.example.org/networkdataset24#> .
@prefix net: <http://purl.org/ontology/network/> .
@prefix prov: <http://www.w3.org/ns/prov#> .
@prefix xsd: <http://www.w3.org/2001/XMLSchema#> .

:BGPUM { :R10.0.3.1 net:inAS :AS10143 . }
:PROVBGP { :BGPUM prov:wasDerivedFrom :BGP_AS10143R1_AS1221R1 .
           }
:METABGP { :PROVBGP prov:generatedAtTime
             "2017-11-02T12:37:00+09:30"^^xsd:dateTime . }

:OSPFLSA { :R10.0.3.1 net:isASBR "true"^^xsd:boolean . }
:PROVLSA { :OSPFLSA prov:wasDerivedFrom :OSPF_AS10143R1 . }
:METALSA { :PROVLSA prov:generatedAtTime
             "2017-11-02T12:37:00+09:30"^^xsd:dateTime . }

:R1CF { :R10.0.3.1 net:hostname "AS10143R1"^^xsd:string . }
:PROVR1CF { :R1CF prov:wasDerivedFrom :AS10143R1_CF . }
:METAR1CF { :PROVR1CF prov:generatedAtTime
              "2017-11-02T12:37:00+09:30"^^xsd:dateTime . }

:OD { :AS10143 net:asName "EXETEL-AS-AP" . }
:PROVOD { :OD prov:wasDerivedFrom :CAIDAASDataset . }
:METAOD { :PROVOD prov:generatedAtTime
            "2017-11-02T12:37:00+09:30"^^xsd:dateTime . }
```

To maximize interoperability, dataset-level provenance can be defined using standard terms from the Vocabulary of Interlinked Datasets (VoID)[22] as shown in Listing 2.12.

Listing 2.12 A dataset whose statements are based on a traceroute

```
@prefix : <http://www.example.com/> .
```

[22]https://www.w3.org/TR/void/

```
@prefix prov: <http://www.w3.org/ns/prov#> .
@prefix rdf: <http://www.w3.org/1999/02/22-rdf-syntax-ns#> .
@prefix void: <http://rdfs.org/ns/void#> .

<http://example.com/networkDiscovery/> a void:dataset ;
prov:wasGeneratedBy :host18_lsa_traceroute_20171106 .
```

This makes it possible to describe datasets and interlink concepts with related LOD datasets efficiently [41], and execute federated queries on LOD datasets [42].

To segregate network knowledge from provenance data and provide provenance for different graphs, partitioned datasets can be declared with different provenance data using `void:subset`. In this context, the `void:Dataset` concept is used strictly for those network knowledge datasets that comply with the LOD dataset requirements.

2.6 Representing Network Data Uncertainty

Network data is inherently uncertain. Probabilistic description logics support the representation of terminological probabilistic knowledge about concepts and roles, and assertional probabilistic knowledge about instances of concepts and roles. For example, probabilistic axioms can be used to express the probability of an IP address being spoofed, or the probability that there are misconfigurations in a network segment that may create vulnerabilities.

In the P-$\mathcal{SROIQ}$ probabilistic description logic [43], for example, the set of individuals I is divided into two disjoint sets, the set of classical individuals I_C, and the finite set of probabilistic individuals I_P, which are associated with every probabilistic individual in the probabilistic ABox. Assume a finite nonempty set C of basic classification concepts, which are atomic or complex concepts C in $\mathcal{SROIQ}^{(\mathcal{D})}$ that are free of individuals from I_P. Every basic classification concept $\phi \in C$ is a classification concept. If ϕ and ψ are classification concepts, then $\neg\phi$ and $(\phi \sqcap \psi)$ are also classification concepts. Note that every classification concept is also a standard $\mathcal{SROIQ}$ concept and conversely, for every finite set S of concepts in $\mathcal{SROIQ}$, there exists a finite set of basic classification concepts such that their set of classification concepts includes S. A conditional statement is an expression of the form $(\psi|\phi)[l, u]$, where ϕ and ψ are classification concepts, and l and u are real numbers from [0, 1]. A probabilistic interpretation Pr is a probability function on I_C that maps every I_C to a value in [0, 1], i.e., Pr: $\mathrm{I}_C \rightarrow [0, 1]$, such that the sum of all Pr(I), where $\mathrm{I} \in \mathrm{I}_C$, is 1. The probability of a classification concept ϕ in a probabilistic interpretation Pr, Pr(ϕ), is the sum of all Pr(I) such that $\mathrm{I} \models \phi$. $\top$ has a satisfying classical interpretation $\mathrm{I} = (\Delta^{\mathcal{I}}, \cdot^{\mathcal{I}})$ iff it has a satisfying probabilistic interpretation Pr.

Possibilistic relationships, such as the possibility that two network devices are identical, or that the traffic between two network devices go through a particular country, can be described with possibilistic description logic axioms. In the π-$\mathcal{SROIQ}$ possibilistic description logic [44], for example, possibilistic concepts represent possibilistic sets of individuals and can be constructed using the rule $C, D \rightarrow \top \mid \bot \mid A \mid C \sqcap D \mid C \sqcup D \mid \neg C \mid \forall R.C \mid \exists R.C \mid \forall T.d \mid \exists T.d \mid \geqslant nS.C \mid \leqslant nS.C \mid \geqslant nT.d \mid \leqslant nT.d \mid \exists S.\text{Self}, (o_i, \alpha_i) \mid (A, \alpha)$, where C, D are (possibly complex) concepts, A denotes atomic concepts, R denotes (possibly complex) abstract roles, S denotes simple roles, T denotes concrete roles, d is a concrete predicate, $n \in \mathbb{N}^+$, and $\alpha \in (0, 1]$. Possibilistic roles represent possibilistic relationships and are built from atomic roles (R_A), inverse roles (R^-), and the universal role U according to the rule $R \rightarrow R_A \mid R^- \mid U$. A possibilistic interpretation $\mathcal{I} = (\Delta^{\mathcal{I}}, \cdot^{\mathcal{I}})$ with respect to the possibilistic concrete domain Δ_D consists of a nonempty set $\Delta^{\mathcal{I}}$, disjoint with $\Delta_{\mathcal{D}}$, (the domain of $\mathcal{I}$) and a function $\cdot^{\mathcal{I}}$, which associated with each concept C a function $C^{\mathcal{I}} : \Delta^{\mathcal{I}} \rightarrow [0, 1]$, with each abstract role R a function $R^{\mathcal{I}} : \Delta^{\mathcal{I}} \times \Delta^{\mathcal{I}} \rightarrow [0, 1]$, with each concrete role T a function $T^{\mathcal{I}} : \Delta^{\mathcal{I}} \times \Delta_{\mathcal{D}} \rightarrow [0, 1]$, with each individual a an element in $\Delta^{\mathcal{I}}$, with each concrete individual v and element in $\Delta_{\mathcal{D}}$, and to each n-ary concrete predicate d the interpretation $d^{\mathcal{D}} : \Delta^n_{\mathcal{D}} \rightarrow [0, 1]$. The semantics of possibilistic concepts are defined as follows:

$$
\begin{aligned}
&\top^{\mathcal{I}}(x) = 1\\
&\bot^{\mathcal{I}}(x) = 0\\
&(C \sqcap D)^{\mathcal{I}}(x) = min\left(C^{\mathcal{I}}(x), D^{\mathcal{I}}(x)\right)\\
&(C \sqcup D)^{\mathcal{I}}(x) = max\left(C^{\mathcal{I}}(x), D^{\mathcal{I}}(x)\right)\\
&(\neg C)^{\mathcal{I}}(x) = 1 - C^{\mathcal{I}}(x)\\
&(\forall R.C)^{\mathcal{I}}(x) = \inf_{y \in \Delta^{\mathcal{I}}} \left\{max\left(1 - R^{\mathcal{I}}(x, y), C^{\mathcal{I}}(y)\right)\right\}\\
&(\exists R.C)^{\mathcal{I}}(x) = \sup_{y \in \Delta^{\mathcal{I}}} \left\{min\left(R^{\mathcal{I}}(x, y), C^{\mathcal{I}}(y)\right)\right\}\\
&(\forall T.d)^{\mathcal{I}}(x) = \inf_{v \in \Delta_{\mathcal{D}}} \left\{max\left(1 - T^{\mathcal{I}}(x, v), d_{\mathcal{D}}(v)\right)\right\}\\
&(\exists T.d)^{\mathcal{I}}(x) = \sup_{v \in \Delta_{\mathcal{D}}} \left\{min\left(T^{\mathcal{I}}(x, v), d_{\mathcal{D}}(v)\right)\right\}\\
&(\geqslant nS.C)^{\mathcal{I}}(x) = \sup_{y_i \in \Delta^{\mathcal{I}}} min_{i=1}^{n} \left\{min\left(S^{\mathcal{I}}(x, y_i), C^{\mathcal{I}}(y_i)\right)\right\}\\
&(\leqslant nS.C)^{\mathcal{I}}(x) = \inf_{y_i \in \Delta^{\mathcal{I}}} max_{i=1}^{n+1} \left\{max\left(1 - S^{\mathcal{I}}(x, y_i), 1 - C^{\mathcal{I}}(y_i)\right)\right\}\\
&(\geqslant nT.d)^{\mathcal{I}}(x) = \sup_{v_i \in \Delta_{\mathcal{D}}} min_{i=1}^{n} \left\{min\left(T^{\mathcal{I}}(x, v_i), d_{\mathcal{D}}(v_i)\right)\right\}\\
&(\leqslant nT.d)^{\mathcal{I}}(x) = \inf_{v_i \in \Delta_{\mathcal{D}}} max_{i=1}^{n+1} \left\{max\left(1 - T^{\mathcal{I}}(x, v_i), 1 - d_{\mathcal{D}}(v_i)\right)\right\}\\
&(\exists S.Self)^{\mathcal{I}}(x) = S^{\mathcal{I}(x,x)}\\
&(\{(o_1, \alpha_1), \ldots, (o_n, \alpha_n)\})^{\mathcal{I}}(x) = sup\left\{\alpha_i / x = o_i^{\mathcal{I}}\right\}\\
&(A, \alpha)^{\mathcal{I}}(x) = A^{\mathcal{I}}(x) \text{ if } A^{\mathcal{I}}(x) \geqslant \alpha, \text{ otherwise } A^{\mathcal{I}}(x) = 0
\end{aligned}
$$

where v represents concrete individuals.

The semantics of possibilistic roles are defined as follows:

$$
\begin{aligned}
&(R_A)^{\mathcal{I}}(x, y) = R_A^{\mathcal{I}}(x, y)\\
&(R^-)^{\mathcal{I}}(x, y) = R^{\mathcal{I}}(y, x)\\
&U^{\mathcal{I}}(x, y) = 1
\end{aligned}
$$

The semantics of possibilistic axioms are defined as follows:

$$
\begin{aligned}
&(C(a))^{\mathcal{I}} = C^{\mathcal{I}}(a^{\mathcal{I}})\\
&(R(a,b))^{\mathcal{I}} = R^{\mathcal{I}}(a^{\mathcal{I}}, b^{\mathcal{I}})\\
&(\neg R(a,b))^{\mathcal{I}} = 1 - R^{\mathcal{I}}(a^{\mathcal{I}}, b^{\mathcal{I}})\\
&(T(a,v))^{\mathcal{I}} = T^{\mathcal{I}}(a^{\mathcal{I}}, vD)\\
&(\neg T(a,v))^{\mathcal{I}} = 1 - T^{\mathcal{I}}(a^{\mathcal{I}}, vD)\\
&(C \sqsubseteq D)^{\mathcal{I}} = \inf_{x\in\Delta^{\mathcal{I}}} \{C^{\mathcal{I}}(x) \Rightarrow D^{\mathcal{I}}(x)\}\\
&(R_1 \dots R_n \sqsubseteq R)^{\mathcal{I}} =\\
&\inf_{x_1,x_{n+1}\in\Delta^{\mathcal{I}}} \left\{sup\{inf(R_1^{\mathcal{I}}(x_1,x_2),\dots,R_n^{\mathcal{I}}(x_n,x_{n+1})) \Rightarrow R_n^{\mathcal{I}}(x_1,x_{n+1})\}\right\}\\
&(T_1 \sqsubseteq T_2)^{\mathcal{I}} = \inf_{x\in\Delta^{\mathcal{I}}, v\in\Delta_{\mathcal{D}}} \left\{T_1^{\mathcal{I}}(x,v) \Rightarrow T_2^{\mathcal{I}}(x,v)\right\}
\end{aligned}
$$

where a and b are abstract individuals.

As a concrete example for representing uncertain network knowledge, consider the triples in Example 2.2.

Example 2.2 Representing uncertain network knowledge
⟨(connectedTo(I10.204.20.14, N10.204.20.12_30))[0.85, 1]⟩
⟨traverses(T10798_36149, Australia), 0.91⟩

The first triple expresses that the probability of connection between interface 10.204.20.14 and network 10.204.20.12/30 is at least 85%, reflecting the accuracy and completeness limitations of network diagnostic tools used to extract this information.

The second triple describes the possibility that the traffic between an AS in South Africa (AS10798) and an AS in Hawaii (AS36149) will go through Australia is 91%.

2.7 Representing Network Data Vagueness

The previous formalisms are related to uncertainty theory, because their statements are either true or false to some probability or possibility. However, when representing cyber-knowledge, not all statements are true or false: many are true to a certain degree [45]. This can be expressed efficiently with fuzzy values taken from a truth space, such as [0, 1] or a complete lattice, which are useful, among other things, to capture imprecise information derived from probe data and vague concepts, such as secure connection and reliable network data.

In the Ł$\mathcal{SROIQ}$ fuzzy description logic [46], for example, fuzzy concepts C and D can be built inductively according to the following syntax rule from atomic concepts (A), top concept ($\top$), bottom concept ($\bot$), named individuals (o_i), and roles R and S, where S is a simple role of the form R^A (atomic role), R^- (negated role), or U (universal role): $C, D \rightarrow A \mid \top \mid \bot \mid C \sqcap D \mid C \sqcup D \mid \neg C \mid \forall R.C \mid \exists R.C \mid \alpha_1/o_1, \dots, \alpha_m/o_m \mid (\geqslant mS.C) \mid (\leqslant nS.C) \mid \exists S.\,\mathrm{Self}$, where $n, m \in \mathbb{N}, n \geqslant 0$, $m > 0$, $o_i \neq o_j$, $1 \leqslant i < j \leqslant m$. Fuzzy general concept inclusions are of the form

$C \sqsubseteq D \geqslant \alpha$ or $C \sqsubseteq D > \beta$, which constrain the truth value of general concept inclusions. Fuzzy assertions can be inequality assertions of the form $a \neq b$, equality assertions of the form $a = b$, and constraints on the truth value of a concept or role assertion of the form $\psi \geqslant \alpha$, $\psi > \beta$, $\psi \leqslant \beta$, or $\psi < \alpha$, where ψ is of the form $C(a)$, $R(a, b)$, or $\neg R(a, b)$. Role axioms can be fuzzy role inclusion axioms of the form $w \sqsubseteq R \geqslant \alpha$ or $w \sqsubseteq R > \beta$ for a role chain $w = R_1 R_2, \ldots, R_n$, or any standard $\mathcal{SROIQ}$ role axioms of the form transitive(R), disjoint(S_1, S_2), reflexive(R), irrefexive(R), symmetric(R), or asymmetric(S). A fuzzy interpretation function $\cdot^{\mathcal{I}}$ in a fuzzy interpretation $\mathcal{I} = (\Delta^{\mathcal{I}}, \cdot^{\mathcal{I}})$ maps each individual name $a \in \mathrm{N}_I$ to an element $a^{\mathcal{I}} \in \Delta^{\mathcal{I}}$, but in contrast to the typical Tarski-style interpretation of crisp DLs, maps each atomic concept $A \in \mathrm{N}_C$ to a membership function $A^{\mathcal{I}} : \Delta^{\mathcal{I}} \rightarrow [0,1]$, rather than subsets of the object domain, and each atomic role $R \in \mathrm{N}_R$ to a membership function $R^{\mathcal{I}} : \Delta^{\mathcal{I}} \times \Delta^{\mathcal{I}} \rightarrow [0, 1]$, rather than subsets of $\Delta^{\mathcal{I}} \times \Delta^{\mathcal{I}}$. Given a t-norm $\otimes$, a t-conorm $\oplus$, a negation function $\ominus$ and an implication function $\Rightarrow$, the fuzzy interpretation is extended to complex concepts and roles as follows:

$\top^{\mathcal{I}}(x) = 1$
$\bot^{\mathcal{I}}(x) = 0$
$(C \sqcap D)^{\mathcal{I}}(x) = C^{\mathcal{I}}(x) \otimes D^{\mathcal{I}}(x)$
$(C \sqcup D)^{\mathcal{I}}(x) = C^{\mathcal{I}}(x) \oplus D^{\mathcal{I}}(x)$
$(\neg C)^{\mathcal{I}}(x) = \ominus C^{\mathcal{I}}(x)$
$(\forall R.C)^{\mathcal{I}}(x) = \inf_{y \in \Delta^{\mathcal{I}}} \left\{ R^{\mathcal{I}}(x, y) \Rightarrow C^{\mathcal{I}}(y) \right\}$
$(\exists R.C)^{\mathcal{I}}(x) = \sup_{y \in \Delta^{\mathcal{I}}} \left\{ R^{\mathcal{I}}(x, y) \otimes C^{\mathcal{I}}(y) \right\}$
$\{\alpha_1/o_1, \ldots, \alpha_m/o_m\}^{\mathcal{I}}(x) = \sup\{\alpha_i | x = o_i^{\mathcal{I}}\}$
$(\geqslant mS.C)^{\mathcal{I}}(x) = \sup_{y_1, \ldots, y_m \in \Delta^{\mathcal{I}}} \left(\min_{i=1}^{m} \left\{ S^{\mathcal{I}}(x, y_i) \otimes C^{\mathcal{I}}(y_i) \right\} \right) \otimes$
$\left(\bigotimes_{1 \leqslant j < k \leqslant m} \left\{ y_j \neq y_k \right\} \right)$
$(\leqslant nS.C)^{\mathcal{I}}(x) = \inf_{y_1, \ldots, y_{n+1} \in \Delta^{\mathcal{I}}} \left(\min_{i=1}^{n+1} \left\{ S^{\mathcal{I}}(x, y_i) \otimes C^{\mathcal{I}}(y_i) \right\} \right) \Rightarrow$
$\left(\bigotimes_{1 \leqslant j < k \leqslant n+1} \left\{ y_j = y_k \right\} \right)$
$(\exists S.\mathrm{Self})^{\mathcal{I}}(x) = S^{\mathcal{I}}(x, x)$
$(R^{-})^{\mathcal{I}}(x, y) = R^{\mathcal{I}}(y, x)$
$U^{\mathcal{I}}(x, y) = 1$

The fuzzy interpretation function is extended to fuzzy axioms as follows:

$(C(a))^{\mathcal{I}} = C^{\mathcal{I}}(a^{\mathcal{I}})$
$R(a, b)^{\mathcal{I}} = R^{\mathcal{I}}(a^{\mathcal{I}}, b^{\mathcal{I}})$
$(\neg R(a, b))^{\mathcal{I}} = \ominus R^{\mathcal{I}}(a^{\mathcal{I}}, b^{\mathcal{I}})$
$(C \sqsubseteq D)^{\mathcal{I}} = \inf_{x \in \Delta^{\mathcal{I}}} \{C^{\mathcal{I}}(x) \Rightarrow D^{\mathcal{I}}(x)\}$

$$(R_1 \ldots R_n \sqsubseteq R)^{\mathcal{I}} = \inf_{x_1, x_{n+1} \in \Delta^{\mathcal{I}}} \left\{ \sup_{x_2 \ldots x_n \in \Delta^{\mathcal{I}}} \{(R_1^{\mathcal{I}}(x_1, x_2) \otimes \ldots \otimes R_n^{\mathcal{I}}(x_n, x_{n+1})) \Rightarrow R^{\mathcal{I}}(x_1, x_{n+1})\} \right\}$$

Using fuzzy axioms, graded propositions can be expressed in the ABox, such as connection between workstation #18 and Node 254 is secure to a degree of 85% (see Example 2.3).

Example 2.3 A fuzzy axiom expressing a graded proposition

$$\mathcal{A} = \left\{ \langle \text{Connection(WKS18\_N254)} \rangle, \langle \text{WKS18\_N254} \sqcap \exists \text{isSecure.}\{\text{true}\} \geqslant 0.85 \rangle \right\}$$

Such statements can be used, for example, to express the difference between connections established over HTTP and HTTPS, or a file transfer via FTP or an encrypted connection using SSL, TLS, SCP, or SFTP.

2.8 Reasoning Support for Cyber-Situational Awareness

The real strength of the formal representation of network knowledge is that it enables inference, thereby making implicit knowledge explicit. Reasoning over the network knowledge also make it possible to perform core tasks, e.g., check whether the knowledge represented in a knowledge base is meaningful by searching for contradictory statements (*KB consistency*), and determine *concept satisfiability*, i.e., check whether a concept can ever have instances. Formally speaking, given knowledge base $\mathcal{K}$ as input, a decision procedure for knowledge base consistency returns "$\mathcal{K}$ is consistent" if there is an interpretation $\mathcal{I}$ such that $\mathcal{I} \models \mathcal{K}$, otherwise it returns "$\mathcal{K}$ is inconsistent." Given concept C and knowledge base $\mathcal{K}$ as input, a decision procedure for concept satisfiability with reference to knowledge base $\mathcal{K}$ returns "C is satisfiable with reference to $\mathcal{K}$" if there is an interpretation $\mathcal{I} = (\Delta^{\mathcal{I}}, \cdot^{\mathcal{I}})$ and an element $d \in \mathcal{I}$ such that $\mathcal{I} \models K$ and $d \in C^{\mathcal{I}}$, otherwise it returns "C is unsatisfiable with reference to $\mathcal{K}$."

Factors that determine the type of reasoning tasks that can be performed on a knowledge base include the level of semantic representation and mathematical formalization, the knowledge base size, presence or absence of entities, and the functions available in the reasoner used for the task.

Reasoning tasks rely on different sets of reasoning rules, such as *RDFS entailment*,[23] which gives semantics to the RDF and RDFS vocabularies; *D-entailment*,[24]

[23]https://www.w3.org/TR/2004/REC-rdf-mt-20040210/#RDFSRules

[24]https://www.w3.org/TR/2004/REC-rdf-mt-20040210/#D_entailment

which provides semantics to datatypes; and the *OWL entailment*,[25] which adds semantics to the OWL vocabulary [47].

For example, based on the triple `:Router10143R1 net:inAS :AS10143` and the domain definition of `net:inAS` in CNTFO (`:inAS rdfs:domain :Router`), it can be inferred using the prp-dom OWL entailment rule, i.e., if T(?p, rdfs:domain, ?c) and T(?x, ?p, ?y), then T(?x, rdf:type, ?c), that `:Router10143R1 a net:Router`, which was not explicitly stated before. Similarly, using the range definition of `net:isAS` (`:inAS rdfs:range :AutonomousSystem`) and the prp-rng OWL entailment rule, i.e., if T(?p, rdfs:range, ?c) and T(?x, ?p, ?y), then T(?y, rdf:type, ?c), that `:AS10143 a net:AutonomousSystem`, which originally was implicit knowledge.

2.9 Conclusions

This chapter described formal knowledge representation formalisms that can be used to capture the semantics of communication network concepts, their properties and the relationships between them, in addition to metadata such as data provenance. Using these state-of-the-art knowledge representation formalisms to reason over network data is a promising research direction for cyber-situational awareness. Semantic Web standards provide syntactic and semantic interoperability for network data derived from diverse sources, and machine interpretable definitions of networking concepts, their properties and relationships, and the entities that instantiate them. The discussed formalisms can employ crisp, fuzzy, probabilistic, or possibilistic logics to express different characteristics of the data, with preference given to those languages which have lower computational complexities, are decidable, and whose axioms can be implemented in OWL. Capturing the semantics of network nodes and routing messages relevant to network topology and information flow can be utilized in reasoning-based network knowledge discovery and ultimately network topology visualization. Expert systems that utilize formally described network semantics can help analysts correlate, combine, and understand network information, disambiguate misleading or conflicting network information, identify vulnerabilities, collaborate with other analysts, and see the larger picture.

References

1. Vishik C, Balduccini M (2015) Making sense of future cybersecurity technologies: using ontologies for multidisciplinary domain analysis. In: Reimer H, Pohlmann N, Schneider W (eds) ISSE 2015. Springer, Wiesbaden, pp 135–145. https://doi.org/10.1007/978-3-658-10934-9_12
2. Sikos LF (2014) Web standards: mastering HTML5, CSS3, and XML, 2nd edn. Apress, New York. https://doi.org/10.1007/978-1-4842-0883-0

[25] https://www.w3.org/TR/owl2-profiles/#Reasoning_in_OWL_2_RL_and_RDF_Graphs_using_Rules

3. Sikos LF (2017) Utilizing multimedia ontologies in video scene interpretation via information fusion and automated reasoning. In: Ganzha M, Maciaszek L, Paprzycki M (eds) Proceedings of the 2017 Federated Conference on Computer Science and Information Systems. IEEE, New York, pp 91–98. https://doi.org/10.15439/2017F66
4. Miksa K, Sabina P, Kasztelnik M (2010) Combining ontologies with domain specific languages: a case study from network configuration software. In: Amann U, Bartho A, Wende C (eds) Reasoning Web: semantic technologies for software engineering. Springer, Heidelberg, pp 99–118. https://doi.org/10.1007/978-3-642-15543-7_4
5. Abar S, Iwaya Y, Abe T, Kinoshita T (2006) Exploiting domain ontologies and intelligent agents: an automated network management support paradigm. In: Chong I, Kawahara K (eds) Information networking: advances in data communications and wireless networks. Springer, Heidelberg, pp 823–832. https://doi.org/10.1007/11919568_82
6. Martínez A, Yannuzzi M, López J, Serral-Gracià R, Ramarez W (2015) Applying information extraction for abstracting and automating CLI-based configuration of network devices in heterogeneous environments. In: Laalaoui Y, Bouguila N (eds) Artificial intelligence applications in information and communication technologies. Springer, Cham, pp 167–193. https://doi.org/10.1007/978-3-319-19833-0_8
7. Quirolgico S, Assis P, Westerinen A, Baskey M, Stokes E (2004) Toward a formal common information model ontology. In: Bussler C, Hong S-k, Jun W, Kaschek R, Kinshuk, Krishnaswamy S, Loke SW, Oberle D, Richards D, Sharma A, Sure Y, Thalheim B (eds) Web information systems–WISE 2004 workshops. Springer, Heidelberg, pp 11–21. https://doi.org/10.1007/978-3-540-30481-4_2
8. Martínez A, Yannuzzi M, Serral-Gracià R, Ramírez W (2014) Ontology-based information extraction from the configuration command line of network routers. In: Prasath R, O'Reilly P, Kathirvalavakumar T (eds) Mining intelligence and knowledge exploration. Springer, Cham, pp 312–322. https://doi.org/10.1007/978-3-319-13817-6_30
9. Laskey K, Chandekar S, Paris B-P (2015) A probabilistic ontology for large-scale IP geolocation. In: Laskey KB, Emmons I, Costa PCG, Oltramari A (eds) Proceedings of the Tenth Conference on Semantic Technology for Intelligence, Defense, and Security. RWTH Aachen University, Aachen, pp 18–25. http://ceur-ws.org/Vol-1523/STIDS_2015_T03_Laskey_etal.pdf
10. ETSI Industry Specification Group (2012) Measurement ontology for IP traffic (MOI); requirements for IP traffic measurement ontologies development. ETSI GS MOI 002 V1.1.1. http://www.etsi.org/deliver/etsi_gs/MOI/001_099/002/01.01.01_60/gs_MOI002v010101p.pdf
11. Kodeswaran P, Kodeswaran SB, Joshi A, Perich F (2008) Utilizing semantic policies for managing BGP route dissemination. In: IEEE INFOCOM workshops 2008. IEEE, New York, pp 184–187. https://doi.org/10.1109/INFOCOM.2008.4544611
12. Voigt S, Howard C, Philp D, Penny C (2018) Representing and reasoning about logical network topologies. In: Croitoru M, Marquis P, Rudolph S, Stapleton G (eds) Graph structures for knowledge representation and reasoning. Springer, Cham, pp 73–83. https://doi.org/10.1007/978-3-319-78102-0_4
13. Sikos LF, Stumptner M, Mayer W, Howard C, Voigt S, Philp D (2018) Representing network knowledge using provenance-aware formalisms for cyber-situational awareness. Procedia Comput Sci 126C:29–38
14. Sikos LF (2016) RDF-powered semantic video annotation tools with concept mapping to Linked Data for next-generation video indexing: a comprehensive review. Multim Tools Appl 76(12):14437–14460. https://doi.org/10.1007/s11042-016-3705-7
15. Bizer C, Heath T, Berners-Lee T (2009) Linked data—the story so far. Int J Semant Web Inform Syst 5(3):1–22. https://doi.org/10.4018/jswis.2009081901
16. Carroll JJ, Bizer C, Hayes P, Stickler P (2005) Named graphs, provenance, and trust. In: Proceedings of the 14th International Conference on World Wide Web. ACM, New York, pp 613–622. https://doi.org/10.1145/1060745.1060835

17. Sikos LF (2017) Description logics in multimedia reasoning. Springer, Cham. https://doi.org/10.1007/978-3-319-54066-5
18. Alani MM (2017) Guide to Cisco routers configuration: becoming a router geek. Springer, Cham. https://doi.org/10.1007/978-3-319-54630-8
19. Systems C (2009) Cisco uBR7200 series universal broadband router software configuration guide. Cisco Press, Indianapolis
20. Rekhter Y, Li T, Hares S (eds) (2006) A border gateway protocol 4 (BGP-4). https://tools.ietf.org/html/rfc4271
21. Moy J (ed) (1998) OSPF version 2. https://tools.ietf.org/html/rfc2328
22. Callon R (ed) (1990) Use of OSI IS-IS for routing in TCP/IP and dual environments. https://tools.ietf.org/html/rfc1195
23. Hedrick C (ed) (1988) Routing information protocol. https://tools.ietf.org/html/rfc1058
24. Nakibly G, Gonikman D, Kirshon A, Boneh D (eds) (2012) Persistent OSPF attacks. In: 19th Annual Network and Distributed System Security Conference, San Diego, CA, USA, 5–8 Feb 2012
25. Dijkstra EW (1959) A note on two problems in connexion with graphs. Numer Math 1(1):269–271. https://doi.org/10.1007/BF01386390
26. Braden R (ed) (1989) Requirements for internet hosts–application and support. https://tools.ietf.org/html/rfc1123
27. Sikos LF, Stumptner M, Mayer W, Howard C, Voigt S, Philp D (2018) Summarizing network information for cyber-situational awareness via cyber-knowledge integration. In: AOC 2018 Convention, Adelaide, Australia, 28–30 May 2018
28. Clemente FJG, Calero JMA, Bernabe JB, Perez JMM, Perez GM, Skarmeta AFG (2011) Semantic Web-based management of routing configurations. J Netw Syst Manag 19(2):209–229. https://doi.org/10.1007/s10922-010-9169-6
29. Udrea O, Recupero DR, Subrahmanian VS (2010) Annotated RDF. ACM Trans Comput Logic 11, Article 10. https://doi.org/10.1145/1656242.1656245
30. Sahoo SS, Bodenreider O, Hitzler P, Sheth A, Thirunarayan K (2010) Provenance context entity (PaCE): scalable provenance tracking for scientific RDF data. In: Gertz M, Ludascher B (eds) Scientific and statistical database management. Springer, Heidelberg, pp 461–470. https://doi.org/10.1007/978-3-642-13818-8_32
31. Nguyen V, Bodenreider O, Sheth A (2014) Don't like RDF reification? Making statements about statements using singleton property. In: Chung C-W (ed) Proceedings of the 23rd International Conference on World Wide Web. ACM, New York, pp 759–770. https://doi.org/10.1145/2566486.2567973
32. Hartig O, Thompson B (2014) Foundations of an alternative approach to reification in RDF. arXiv:1406.3399
33. Zimmermann A, Gimenez-Garcea JM (2017) Integrating context of statements within description logics. arXiv:1709.04970
34. Watkins ER, Nicole DA (2006) Named graphs as a mechanism for reasoning about provenance. In: Zhou X, Li J, Shen HT, Kitsuregawa M, Zhang Y (eds) Frontiers of WWW research and development. Springer, Heidelberg, pp 943–948. https://doi.org/10.1007/11610113_99
35. Flouris G, Fundulaki I, Pediaditis P, Theoharis Y, Christophides V (2009) Coloring RDF triples to capture provenance. In: Bernstein A, Karger DR, Heath T, Feigenbaum L, Maynard D, Motta E, Thirunarayan K (eds) The Semantic Web–ISWC 2009. Springer, Heidelberg, pp 196–212. https://doi.org/10.1007/978-3-642-04930-9_13
36. Pediaditis P, Flouris G, Fundulaki I, Christophides V (2009) On explicit provenance management in RDF/S graphs. In: Proceedings of the First Workshop on the Theory and Practice of Provenance, Article 4. USENIX Association, Berkeley
37. Groth P, Gibson A, Velterop J (2010) The anatomy of a nanopublication. Inform Serv Use 30(1–2):51–56. https://doi.org/10.3233/ISU-2010-0613
38. Straccia U, Lopes N, Lukácsy G, Polleres A (2010) A general framework for representing and reasoning with annotated semantic web data. In: Proceedings of the 24th AAAI Conference on Artificial Intelligence. AAAI Press, Menlo Park, CA, USA, pp 1437–1442. https://www.aaai.org/ocs/index.php/AAAI/AAAI10/paper/view/1590/2228

39. Schüler B, Sizov S, Staab S, Tran DT (2008) Querying for meta knowledge. In: Proceedings of the 17th International Conference on World Wide Web. ACM, New York, pp 625–634. https://doi.org/10.1145/1367497.1367582
40. Sikos LF (2015) Mastering structured data on the Semantic Web: from HTML5 Microdata to Linked Open Data. Apress, New York. https://doi.org/10.1007/978-1-4842-1049-9
41. Alexander K, Cyganiak R, Hausenblas M, Zhao J (2009) Describing linked datasets. In: Bizer C, Heath T, Berners-Lee T, Idehen K (eds) Proceedings of the WWW2009 Workshop on Linked Data on the Web. RWTH Aachen University, Aachen. http://ceur-ws.org/Vol-538/ldow2009_paper20.pdf
42. Akar Z, Halaç TG, Ekinci EE, Dikenelli O (2012) Querying the Web of interlinked datasets using VoID descriptions. In: Bizer C, Heath T, Berners-Lee T, Hausenblas M (eds) Proceedings of the WWW2012 Workshop on Linked Data on the Web. RWTH Aachen University, Aachen. http://ceur-ws.org/Vol-937/ldow2012-paper-06.pdf
43. Klinov P, Parsia B (2013) Understanding a probabilistic description logic via connections to first-order logic of probability. In: Bobillo F, Costa PCG, d'Amato C, Fanizzi N, Laskey KB, Laskey KJ, Lukasiewicz T, Nickles M, Pool M (eds) Uncertainty reasoning for the Semantic Web II. Springer, Heidelberg, pp 41–58. https://doi.org/10.1007/978-3-642-35975-0_3
44. Bal-Bourai S, Mokhtari A (2016) π-$\mathcal{SROIQ}^{(\mathcal{D})}$: possibilistic description logic for uncertain geographic information. In: Fujita H, Ali M, Selamat A, Sasaki J, Kurematsu M (eds) Trends in applied knowledge-based systems and data science. Springer, Cham, pp 818–829. https://doi.org/10.1007/978-3-319-42007-3_69
45. Sikos LF (2018) Handling uncertainty and vagueness in network knowledge representation for cyberthreat intelligence. In: Proceedings of the 2018 IEEE International Conference on Fuzzy Systems. Curran Associates, Red Hook, NY, USA
46. Bobillo F, Straccia U (2011) Reasoning with the finitely many-valued Łukasiewicz fuzzy description logic $\mathcal{SROIQ}$. Inform Sci 181(4):758–778. https://doi.org/10.1016/j.ins.2010.10.020
47. Sikos LF, Stumptner M, Mayer W, Howard C, Voigt S, Philp D (2018) Automated reasoning over provenance-aware communication network knowledge in support of cyber-situational awareness. In: Liu W, Giunchiglia F, Yang B (eds) Knowledge science, engineering, and management. Springer, Cham, pp 132–143. https://doi.org/10.1007/978-3-319-99247-1_12

Chapter 3
The Security of Machine Learning Systems

Luis Muñoz-González and Emil C. Lupu

Abstract Machine learning lies at the core of many modern applications, extracting valuable information from data acquired from numerous sources. It has produced a disruptive change in society, providing new functionality, improved quality of life for users, e.g., through personalization, optimized use of resources, and the automation of many processes. However, machine learning systems can themselves be the targets of attackers, who might gain a significant advantage by exploiting the vulnerabilities of learning algorithms. Such attacks have already been reported in the wild in different application domains. This chapter describes the mechanisms that allow attackers to compromise machine learning systems by injecting malicious data or exploiting the algorithms' weaknesses and blind spots. Furthermore, mechanisms that can help mitigate the effect of such attacks are also explained, along with the challenges of designing more secure machine learning systems.

3.1 Machine Learning Algorithms Are Vulnerable

Advances in *machine learning*[1] have produce a disruptive change in the society and the development of new technologies. In the *Big Data*[2] era, an increasing number of services rely on AI and data-driven approaches that leverage the huge amount of data available from diverse sources, including people, devices, and sensors. Machine learning algorithms allow to extract valuable information from this overwhelming amount of data, providing powerful predictive capabilities. The use of machine learning facilitates the automation of many tasks, and brings important benefits in terms of new functionality, personalization, and optimization of resources.

[1]Machine learning is a field of computer science that gives software tools the ability to progressively improve their performance on a specific task without being explicitly programmed.

[2]Big data refers to extremely large datasets that, when analyzed, can reveal patterns, trends, and associations, but cannot be processed with traditional data processing tools due to data velocity, volume, value, variety, and veracity.

L. Muñoz-González (✉) · E. C. Lupu
Imperial College London, London, UK
e-mail: l.munoz@imperial.ac.uk

© Springer Nature Switzerland AG 2019

L. F. Sikos (ed.), *AI in Cybersecurity*, Intelligent Systems Reference Library 151,
https://doi.org/10.1007/978-3-319-98842-9_3

Machine learning has been successfully applied in many different application domains, including computer and system security. Thus machine learning is at the core of most non-signature-based detection systems, including, among other things, spam, malware, network intrusions, and fraudulent activities. In contrast to traditional signature-based systems, machine learning has generalization capabilities, i.e., learning algorithms can produce predictions for samples they have not seen before.

Despite the benefits of machine learning technologies, learning algorithms can be abused, providing new opportunities to cyber-criminals to conduct illicit and highly profitable activities. It has been shown that machine learning algorithms are vulnerable and can be the objectives of attackers, who might gain a significant benefit by exploiting the vulnerabilities of these algorithms [1, 2]. In fact, machine learning itself can be the *weakest link* in the security chain, and its vulnerabilities can be exploited by attackers to compromise entire infrastructures. Attackers can inject malicious data to poison the learning process or manipulate data at test time, exploiting the blind spots and weaknesses of the learning algorithm to evade detection.

Far from a merely theoretical hypothesis, these attacks have already been reported in real-world systems, such as antivirus engines, spam filters, and fake profile and fake news detection [1]. These attacks have fostered the investigation of the security properties of machine learning. Therefore, there is a growing interest in *adversarial machine learning* [3], a research area that lies at the intersection of machine learning and cybersecurity, which aims to understand the vulnerabilities of existing machine learning algorithms and the development of new, more secure algorithms.

This chapter describes the vulnerabilities of machine learning systems, explaining the mechanisms that allow attackers to compromise the learning algorithms at training and test time, and are commonly referred to as *poisoning* and *evasion attacks*, respectively. First, the threat models describing the possible attack scenarios depending on the attacker's capability, goal, knowledge, and strategy are formalized. Then, poisoning attacks are described, demonstrating how attackers can manipulate machine learning systems by injecting malicious data into the training dataset used by the algorithm to learn the parameters. Some defensive techniques are provided that can help mitigate the effect of such attacks. Moreover, evasion attacks are also described, which are attacks produced at test time and aim to produce intentional errors or evade the system. Not only attack strategies are described, but also some defense strategies that can help reduce their success. However, defense against poisoning and evasion attacks still remains an open research problem.

3.2 Threat Model

Similar to other traditional security contexts, the first step towards the analysis of the vulnerabilities and the security aspects of a system requires the definition of an adequate threat model. In this section, the framework originally proposed by Barreno et al. [4, 5] and Huang et al. [3] is described with the extensions of Biggio et al. [6] and Muñoz-González et al. [7], which encompasses different attack scenarios against machine learning algorithms with different attack models. The framework characterizes the attacks according to the attacker's goal, his capability to manipulate the data and influence the learning system, his familiarity with the algorithms, the data used by the defender, and the attacker's strategy. These aspects allow the definition of optimal attack strategies as an optimization problem, whose solution provides the construction of attack samples.

This framework is valid for attacks both at training and test time, usually referred to as poisoning and evasion attacks [3, 6]. When the attack samples target *deep learning* algorithms, these samples are also known as *adversarial (training and test) samples* [7–11]. Although an important part of the related work in adversarial machine learning focuses on classification tasks, this framework can also be used for the description of threats for other learning tasks and machine learning paradigms (such as *unsupervised learning*).[3]

3.2.1 Threats by the Capability of the Attacker

The capability of attackers to compromise a machine learning system can be defined based on the *influence* the attackers have on the data used by the learning algorithm and on the presence of *data manipulation constraints*.

Attack Influence

According to the first aspect, the attack influence can be *causative*, if the attacker can inject or manipulate the training data used by the learning algorithm, or *exploratory*, if the attacker cannot influence the training process, but can create malicious samples at test time (poisoning and evasion attacks).

Poisoning attacks occur when the data collected to train the learning algorithms is untrusted, i.e., data is collected and labeled from sensors, people, or devices that can be compromised or maliciously manipulated. In these applications, given the huge amount of data collected, manual data curation to eliminate malicious examples and outliers is often not feasible. Moreover, many applications require frequent retraining of the learning algorithms to adapt to changes in the underlying data distribution.

In exploratory attacks, the attacker can only manipulate data at test time. Even if the dataset used for training the learning algorithm is trusted, the attacker can probe the system to learn the weaknesses and blind spots to produce intentional errors in the learning system (evasion attacks). However, exploratory attacks can also include scenarios in which the objective of the attacker is to obtain information about the machine learning model used by the defender or the data used for training, which in either case means privacy violation.

[3]Unsupervised machine learning refers to machine learning tasks that infer a function to describe hidden structure from unlabeled data.

Data Manipulation Constraints

Another aspect regarding the attacker's capabilities is the potential presence of constraints for data manipulation to carry out attacks. This strongly depends on the application domain. In computer vision applications, for example, the attacker might be able to manipulate every pixel in the image with the only constrain of providing valid values for each pixel. In contrast, if an attacker aims to evade a malware classification system, the manipulation of the data must preserve the malicious functionality of the malware, limiting the degrees of freedom to carry out an attack.

In poisoning attacks, the attacker may not be in control of the labels assigned to the injected poisoning points. For example, in spam detection systems, the labels are usually assigned automatically, and then (sometimes) relabeled with user feedback. However, an attacker can infer the labels that are likely to be assigned to the poisoning points. This way, the attacker attempts to control the maximum amount of perturbation added to the input data to have the attack points labeled as desired. This may also be important to generate poisoning samples that are more difficult to detect with outlier detection and data prefiltering, which can mitigate the effect of some attacks [12].

Typically, these constraints can be modeled in the definition of optimal attack strategies, and can be characterized by assuming that an initial set of attack samples $\mathcal{D}_p$ is given and that the attack points are modified according to a space of possible modifications $\phi(\mathcal{D}_p)$, such as by constraining the norm of the input perturbation of each attack point.

3.2.2 Threats by the Goal of the Attacker

The goal of the attack can be described in terms of the desired *security violation* and *attack specificity*. In some tasks, such as in multi-class classification scenarios, the attacker's goal must also be described in terms of *error specificity*, i.e., the type of errors the attacker aims to produce in the system [7].

Security Violation

In our context, security violation refers to high-level security violations caused by attacks. Three different security violations can be distinguished against machine learning systems:

- *Integrity violation*: the attack evades detection without compromising the normal operation of the system. For example, in a spam filter application this can be achieved by generating spam emails that are misclassified as solicited emails (produce false negatives).
- *Availability violation*: the attacker aims to compromise system functionality, as for example, by increasing the overall error rate in a classification algorithm.
- *Privacy violation*: the attacker obtains private information about the machine learning system, the data used for training, or the users of the system.

Attack Specificity

Attack specificity defines how specific the attacker's intention is, representing a continuous spectrum of possibilities ranging from targeted to nondiscriminatory scenarios:

- *Targeted attack*: the attacker aims to produce errors or to degrade system performance for a reduced set of samples.
- *Indiscriminate attack*: the attacker aims to degrade the performance of the system or to produce errors in a nondiscriminatory fashion (for a broad set of samples).

Error Specificity

Error specificity disambiguates cases in which the nature of the errors can be different, as in multi-class classification tasks [7]. Therefore, from the error specificity point of view, an attack can be one of the following:

- *Error-agnostic attack*: the attacker aims to produce any kind of error in the system. For example, in a computer vision application, an error-agnostic attack is produced when an attacker modifies an image so that the machine learning system misclassifies the object represented in that image, regardless of the (incorrect) predicted category.
- *Error-specific attack*: the attacker aims to produce a specific type of error. In a computer vision application, for example, this is the case if the attacker manages the malicious image to be misclassified by the machine learning system as a specific (incorrect) class, say, if the attacker aims to misclassify the picture of a dog as a cat.

3.2.3 Threats by the Knowledge of the Attacker

Attacker may have various levels of knowledge about the targeted machine learning system, including the following:

- The dataset used for training the learning algorithm, $\mathcal{D}_{tr}$.
- The set of features used by the learning algorithm, $\mathcal{X}$.
- The learning algorithm, $\mathcal{M}$.
- The objective function the learning algorithm aims to optimize, $\mathcal{L}$.
- The parameters of the machine learning algorithm, $\boldsymbol{w}$.

Thus, attacker knowledge can be described in terms of a vector $\boldsymbol{\theta} = (\mathcal{D}_{tr}, \mathcal{X}, \mathcal{M}, \mathcal{L}, \boldsymbol{w})$ that encodes the aforementioned elements of the machine learning system. This representation encompasses many different attack scenarios, depending on the assumptions made on each component of $\boldsymbol{\theta}$. Typically, only two main settings are considered though, namely *perfect* and *limited* knowledge attacks.

Perfect Knowledge Attacks

In the case of perfect knowledge attacks, the attacker is assumed to know everything about the targeted system. Although this may be unrealistic in most cases, perfect knowledge attacks allow to perform worst-case evaluations of the security of machine learning systems. This enables the estimation of upper bounds on the performance degradation of a system under attack. It can also be helpful for model selection, comparing the performance of different learning algorithms to different settings while taking into account the possibility of being attacked.

Limited Knowledge Attacks

Limited knowledge attacks include a broad range of possibilities, but the literature typically considers two cases: attacks with *surrogate data* and attacks with *surrogate learners*.

- *Limited knowledge attacks with surrogate data* are attacks in which the attacker knows the feature representation $\mathcal{X}$, the learning algorithm $\mathcal{M}$, and the objective function optimized by the learning algorithm, $\mathcal{L}$. These attacks also assume that the attacker does not have access to the training data $\mathcal{D}_{tr}$, but that she has access to a surrogate dataset, $\hat{\mathcal{D}}_{tr}$, with similar characteristics and data distribution than $\mathcal{D}_{tr}$. Then, the attacker estimates the parameters of the learning algorithm $\hat{\boldsymbol{w}}$ by optimizing the objective function $\mathcal{L}$ on the surrogate dataset $\hat{\mathcal{D}}_{tr}$.
- In *limited knowledge attacks with surrogate models*, the attacker is assumed to know the training data $\mathcal{D}_{tr}$ and the feature set $\mathcal{X}$ (e.g., if the learning algorithm is trained on publicly available datasets), but not the learning algorithm and the optimized objective function $\mathcal{L}$. In this case, the estimated parameter vector $\hat{\boldsymbol{w}}$ may also belong to a different vector space than that of the targeted system, as the model of the attacker can be different. These attacks also include cases in which, regardless whether the attacker knows the learning algorithm of the targeted system, it is not possible for the attacker to derive an optimal attack strategy. This happens if the corresponding optimization problem is intractable or difficult to solve, in which case a surrogate model can help overcome this difficulty by carrying out a tractable but (possibly) less effective attack.

3.2.4 Threats by Attack Strategy

The attack strategy can be formulated as an optimization problem that considers the different aspects of the threat model. Thus, given the attacker's knowledge $\boldsymbol{\theta}$ and a set of malicious samples $\mathcal{D}_p \in \phi(\mathcal{D}_p)$ the attacker wants to produce, the attacker's goal can be characterized in terms of an objective function $\mathcal{A}(\mathcal{D}_p, \boldsymbol{\theta}) \in \mathcal{R}$, which measures attack efficiency with respect to the attack points $\mathcal{D}_p$. Based on these, the optimal attack strategy can be written as the following optimization problem:

$$\mathcal{D}_p^* \in \arg\max_{\mathcal{D}_p \in \phi(\mathcal{D}_p)} \mathcal{A}(\mathcal{D}_p, \boldsymbol{\theta}) \tag{3.1}$$

This high-level formulation includes both poisoning and evasion attacks, and can be applied to different learning tasks. However, the rest of the chapter focuses on classification tasks of adversarial machine learning. Finally, Table 3.1 summarizes the different aspects of the described threat model.

Table 3.1 Threat model

Attacker's capability	**Attack influence**: • *Causative attacks*: the attacker can manipulate the training data and influence the learning algorithm • *Exploratory attacks*: the attacker can only manipulate data at test time
	Data manipulation constraints: depending on the application, the attacker may be limited to manipulate the features or the labels of the attack samples. The attacker may also consider additional constraints to avoid detection.
Attacker's goal	**Security violation**: • *Integrity violation*: malicious activities do not compromise normal system operation • *Availability violation*: normal system operation is compromised • *Privacy violation*: the attacker obtains private information about the learning algorithm, the data or the users of the system
	Attack specificity: • *Targeted attack*: focused on a reduced number of target samples • *Indiscriminate attack*: targets a broad range of samples, e.g., by increasing the overall error rate of the learning system
	Error specificity: • *Error-agnostic attack*: the attacker aims to produce errors in the system regardless of the kind of error • *Error-specific attack*: the attacker aims to produce specific types of errors in the system
Attacker's knowledge	**Perfect knowledge**: the attacker is assumed to know the dataset and the feature set used to train the learning algorithm, the model, the function optimized to train the algorithm, and its parameters
	Limited knowledge: • *Surrogate data*: the attacker does not know the training dataset used in the target system, but has access to a surrogate dataset with similar characteristics and data distribution • *Surrogate model*: the attacker knows the dataset used by the target system, but does not know anything about the target model and the function to be optimized. The attacker uses a surrogate model to craft the attack points.

3.3 Data Poisoning

Data poisoning (also known as *causative attacks*) are considered among the most relevant and emerging security threats for data-driven technologies [13]. In these attacks, the attacker is assumed to have some control over a fraction of the training data used by the learning algorithm. The goal of the attacker is to subvert the entire learning process in a nondiscriminatory or targeted way, i.e., aiming to decrease the overall performance of the system or to produce particular kinds of errors. These attacks can also facilitate subsequent evasion of the system.

In many applications, data is collected in the wild from untrusted sources, such as sensors, devices, or information from humans. For example, many systems and *online* services rely on users' data and feedback on their decisions to train and update the underlying learning algorithms. This threat is part of a more general problem related to the reliability of the large amount of data collected by these systems. Obviously, an attacker could provide information purposefully generated to gradually poison the system and compromise its performance over time. Even if the malicious data injected by the attacker is correctly classified or identified by the learning algorithm, its performance is still degraded when using this poisoned data to retrain the model. For instance, in the context of spam filtering applications, machine learning algorithms usually classify an email as *spam* or *ham* (i.e., *solicited email*) based on, among other features, the words contained in the body and the header of the email. Attackers can send malicious emails by mixing spam and ham words. As a result, when the system is retrained to adapt to new forms of spam, some of the words previously considered by the system as indications of solicited email will now be considered as typical spam. In addition, some emails containing legitimate words will be incorrectly classified as spam.

In a classification task, the attacker aims to modify the *decision boundary*[4] learned by the machine learning algorithm to increase the overall classification error or the error rate of some specific classes, or, alternatively, to produce misclassification for specific sets of data points. Following the previous example for spam filtering applications, the attacker may want to make the system unusable by having both spam and ham emails classified incorrectly, or by increasing the number of false negatives, i.e., spam classified as ham. The attacker can also target some specific emails that she wants to be misclassified as spam, for example targeting advertising emails from a competitor [14].

Figure 3.1 shows an example of a poisoning attack for a binary classifier with a small *neural network*,[5] where the system aims to correctly classify the blue and the red dots, assuming that the attacker can only inject malicious points (labeled as red). The decision boundary before poisoning is depicted with a blue line. Figure 3.1a shows that after injecting a single poisoning point (represented as a red star), the decision

---

[4] A decision boundary is a hypersurface that partitions the underlying vector space into multiple sets, one for each class.

[5] Artificial neural networks are computing systems inspired by the biological neural networks of brains.

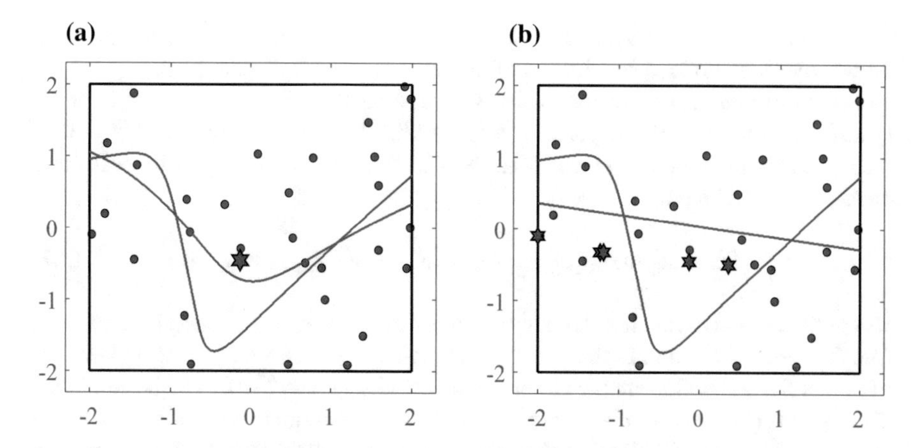

Fig. 3.1 Example of data poisoning in a binary classification problem injecting **a** one poisoning point and **b** five poisoning points. The blue line depicts the decision boundary when the system is not attacked. The decision boundary after poisoning is depicted in red. The poisoning points belong to the red class and are represented with stars

boundary (red line) changes noticeably. In Fig. 3.1b represents the significant effect of the attack after adding five malicious points.

In this section, different poisoning attack scenarios against machine learning classifiers are reviewed, using the threat model described in Sect. 3.2. In addition, a formulation is described for optimal poisoning attack strategies, which allows us to model worst-case scenarios that can be applied to assess the security of machine learning classifiers when considering the possibility of data poisoning. The transferability property of poisoning attacks is also explained, meaning that attacks against a particular learning algorithm are usually effective against other learning algorithms for the same task. Finally, mechanisms that can help to mitigate the effect of these attacks are described.

3.3.1 Poisoning Attack Scenarios

Following a treatment similar to [7], two main scenarios are described that cover most of the possible attacks against multi-class classification systems, according to the type of errors the attacker aims to cause.

Error-Agnostic Poisoning Attacks

Error-agnostic poisoning attacks are the most common poisoning attacks [15–17], in which binary classification tasks are typically considered only to cause a *Denial of Service* (DoS) attack. In these attacks, the attacker aims to produce errors in the system, but it does not matter what type of errors. From the security violation point of view, these are availability attacks, which can be targeted indiscriminately depending on whether they affect a specific set of data points or a single data point.

For multi-class classification systems, error-agnostic attacks do not aim to cause specific errors, but only to produce misclassification on the targeted data points. Using the formulation of the attack strategy in Eq. 3.1, an error-agnostic poisoning attack can be written as

$$\mathcal{D}_p^* \in \arg\max_{\mathcal{D}_p \in \phi(\mathcal{D}_p)} \mathcal{A}(\mathcal{D}_p, \boldsymbol{\theta}) = \mathcal{L}(\mathcal{D}_{target}, \boldsymbol{w}(\mathcal{D}_p)) \quad (3.2)$$

where the objective function the attacker aims to maximize is defined in terms of a *loss function*[6] $\mathcal{L}$ (similar to the one used by the targeted classifier) evaluated in a set of points $\mathcal{D}_{target}$, which are the targets of the attacker. Thus, $\mathcal{D}_{target}$ can be a reduced set of points (in the case of a targeted attack), or a representative set of points drawn from a similar data distribution than the one used by the victim (in the case of an indiscriminate attack).

Loss $\mathcal{L}$ is also a function of the parameters of the learning algorithm, $\boldsymbol{w}$, which depends on the set of poisoning points injected, $\mathcal{D}_p$, i.e., the dependency of $\mathcal{L}$ on $\mathcal{D}_p$ is indirectly encoded through the parameters $\boldsymbol{w}$. As described later in Sect. 3.3.2, this implicit dependency can be modeled as a *bilevel optimization*[7] problem.

Therefore, in error-agnostic poisoning attacks the attacker wants to maximize the loss function for the targeted samples, regardless of the classification errors that can be produced in the system. This scenario allows to model poisoning attacks in which the attacker can manipulate both the features and the labels of the malicious points.

Error-Specific Poisoning Attacks

In error-specific poisoning attacks, it is assumed that the attacker wants to cause specific misclassification errors, which is relevant in multi-class classification tasks [7]. According to the taxonomy described in Sect. 3.2, this kind of attack can produce both integrity and availability security violations, and, as in the previous case, it can be targeted or nondiscriminatory, depending on the set of data points targeted by the attacker.

The attack strategy for error-specific poisoning attacks can be formulated as

$$\mathcal{D}_p^* \in \arg\max_{\mathcal{D}_p \in \phi(\mathcal{D}_p)} \mathcal{A}(\mathcal{D}_p, \boldsymbol{\theta}) = \arg\max_{\mathcal{D}_p \in \phi(\mathcal{D}_p)} - \mathcal{L}(\mathcal{D}'_{target}, \boldsymbol{w}(\mathcal{D}_p)) \quad (3.3)$$

or, equivalently

$$\mathcal{D}_p^* \in \arg\min_{\mathcal{D}_p \in \phi(\mathcal{D}_p)} \mathcal{L}(\mathcal{D}'_{target}, \boldsymbol{w}(\mathcal{D}_p)) \quad (3.4)$$

In this case, the set of target samples $\mathcal{D}'_{target}$ contains the same data as $\mathcal{D}_{target}$, but with labels chosen by the attacker. For example, if the attacker wants to misclassify a sample from class 1 (in the dataset $\mathcal{D}_{target}$) as if it was a sample from class 2,

[6]A loss function is a function that maps values of one or more variables onto a real number representing the cost associated with those values.

[7]Bilevel optimization is an optimization that embeds (nests) a problem within another problem.

the attacker must label that sample in $\mathcal{D}'_{target}$ as class 2. Then, the objective of the attacker is to find the set of poisoning points, $\mathcal{D}_p$, that minimize the loss function on the relabeled samples in $\mathcal{D}'_{target}$. As in the previous case, the dependency of the objective function w.r.t. $\mathcal{D}_p$ is indirect (through the parameters of the model $\boldsymbol{w}$). This can also be modeled as a bilevel optimization problem.

For integrity violations and targeted attacks, some of the labels in $\mathcal{D}'_{target}$ may actually be the same as the true labels, so that normal system operation is not compromised, or only specific data points are affected. This can also help to make the attack less detectable.

Synthetic Example

To illustrate the previous scenarios, Fig. 3.2 shows a synthetic dataset for multi-class classification with three classes, represented as blue, green, and red dots. Figure 3.2a suggests that when there is no attack, the multi-class logistic regression classifier used in this example is capable of separating the three classes correctly (the decision regions are depicted with the same colors as the corresponding data points). For the attacks, it is assumed that the attacker can only inject data points labeled as green.

Figure 3.2b shows the resulting decision regions after an error-agnostic poisoning attack, during which the attacker injected three malicious points. These points caused significant changes in the decision boundary, compared to what can be learned when no attack is present (depicted with a blue line). The poisoning points are represented as green dots with a red border and are situated at the corners of a black rectangle that represents the attacker's constraints to manipulate the features (i.e., $\phi(\mathcal{D}_p)$ in Eq. 3.2). In this case, there are two overlapping poisoning points in the upper-left corner.

Figure 3.2c shows the effect of an error-specific poisoning attack, for which the attacker wants to have the red points misclassified as green. After injecting three poisoning points, the decision region for the red class have decreased accordingly. As a consequence, most of the red points will now be misclassified as green. As in Fig. 3.2c, the decision boundary learned for the clean dataset is represented with a blue line and the black rectangle represents the attacker's constraints in Eqs. 3.3 and 3.4.

3.3.2 *Optimal Poisoning Attacks*

The optimal formulations of the poisoning attack strategies described previously can help to understand the vulnerabilities of machine learning classifiers and to assess the robustness of the algorithms against data poisoning in worst-case scenarios. As mentioned before, the two poisoning attack scenarios described in Eqs. 3.2 and 3.3 can be modeled as bilevel optimization problems.

In line with most related work in poisoning attacks [7, 15–17], the following assumptions can be made to reduce the complexity of the problem:

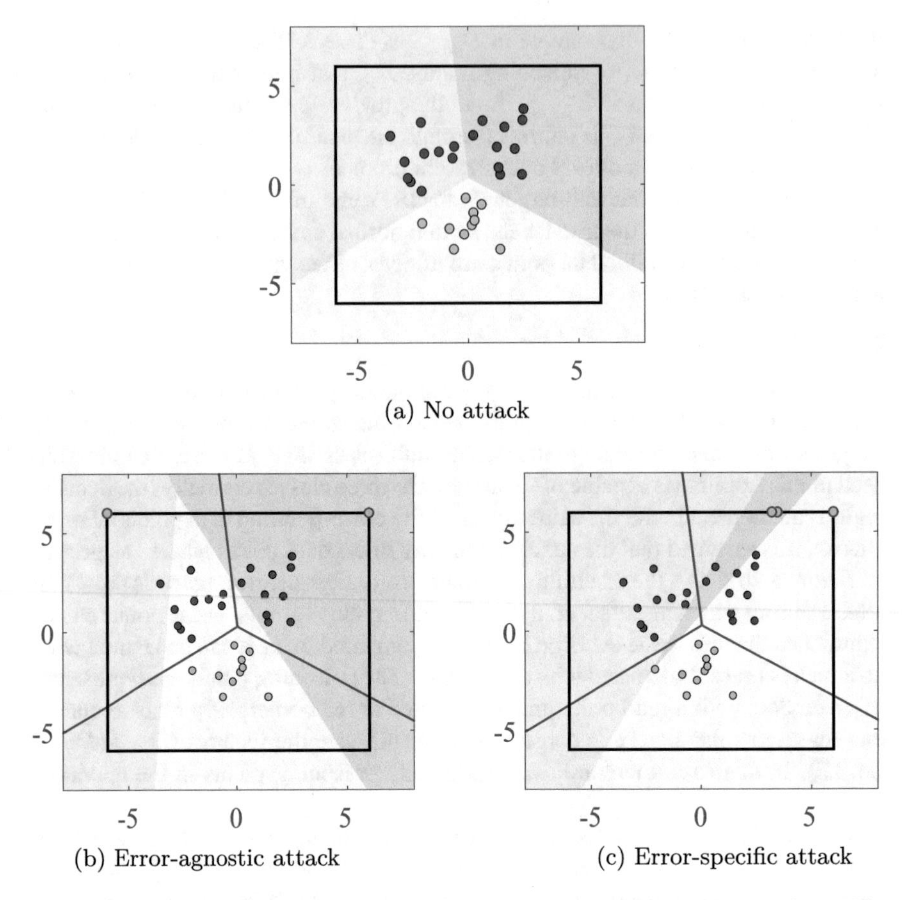

Fig. 3.2 Poisoning attack scenarios in a multi-class classification problem with three classes. The objective of the error-agnostic attack in (**b**) is to increase the overall classification error. The error-specific attack in (**c**) aims to misclassify red points as green. The poisoning points belong to the green class and are denoted with a red border. The blue line represents the decision boundary before the attack and the black rectangle represents the attacker's constraints to generate the malicious data points

- Consider the optimization of one poisoning point at a time.
- Assume that the labels of the poisoning points are chosen initially by the attacker, and then kept fixed during the optimization of the features of the poisoning points.

With these assumptions, optimal poisoning attacks can be simplified as

$$\begin{aligned} \boldsymbol{x}_p^* &\in \arg\max_{\boldsymbol{x}_p \in \phi(\boldsymbol{x}_p)} \mathcal{A}(\mathcal{D}_p, \boldsymbol{\theta}) = \arg\max_{\boldsymbol{x}_p \in \phi(\boldsymbol{x}_p)} \mathcal{L}(\mathcal{D}_{target}, \boldsymbol{w}(\mathcal{D}_p)) \,, \\ &\text{s.t.} \;\; \mathbf{w}^* \in \arg\min_{\mathbf{w}} \; \mathcal{L}_{tr}(\mathcal{D}_{tr} \cup (\boldsymbol{x}_p, y_p), \mathbf{w}) \end{aligned} \tag{3.5}$$

where $\mathcal{D}_{tr}$ is the training dataset used by the learning algorithm and the pair $(\boldsymbol{x}_p, y_p)$ represents the features and the label of the poisoning point, respectively. Thus, the outer optimization problem in Eq. 3.5 represents the attacker's problem, whereas the inner problem is related to the defender's objective, in other words, to minimize the loss function, $\mathcal{L}_{tr}$, on the (tainted) training dataset. Usually $\mathcal{L}$ and $\mathcal{L}_{tr}$ are expected to be the same (in perfect knowledge scenarios), but by using a different notation for the attacker's and defender's losses, more flexible scenarios can be considered.

This bilevel optimization problem can be solved with a gradient ascent approach for some classes of loss functions and learning algorithms, for which the gradients can be computed. This is the case of, among other things, neural networks, deep learning architectures, and support vector machines (SVM). However, it is still not clear how to solve this problem for decision trees or random forests, to which a gradient-based approaches cannot be applied.

Provided that $\mathcal{L}$ and $\mathcal{L}_{tr}$, the loss for the attacker and the defender, are differentiable w.r.t. $\boldsymbol{w}$ and $\boldsymbol{x}_p$, the gradient $\nabla_{\boldsymbol{x}_p}\mathcal{A}$ can be computed, which is needed to solve the bilevel optimization problem, following the chain rule:

$$\nabla_{\boldsymbol{x}_p}\mathcal{A} = \nabla_{\boldsymbol{x}_p}\mathcal{L} + \frac{\partial \boldsymbol{w}}{\partial \boldsymbol{x}_p}\nabla_{\boldsymbol{w}}\mathcal{L} \tag{3.6}$$

The main difficulty here is to compute the implicit derivative of the parameters $\boldsymbol{w}$ w.r.t. the poisoning point $\boldsymbol{x}_p$. To do so, the inner optimization problem in Eq. 3.5 can be replaced with its stationarity conditions, i.e., the *Karush-Kuhn-Tucker* (KKT) conditions [7, 15–17]:

$$\nabla_{\boldsymbol{w}}\mathcal{L}_{tr}(\mathcal{D}_{tr} \cup (\boldsymbol{x}_p, y_p), \mathbf{w}) = \mathbf{0} \tag{3.7}$$

This can only be done under some regularity conditions, requiring $\mathcal{L}_{tr}$, the loss function of the learning algorithm, to be differentiable w.r.t. $\boldsymbol{w}$ and $\boldsymbol{x}_p$. This is the case of linear classifiers, such as *Adaline*, support vector machines, or logistic regression, neural networks, and deep learning architectures.

Following from the KKT conditions in Eq. 3.7, the *implicit function theorem* can be applied to compute the desired implicit derivative:

$$\frac{\partial \boldsymbol{w}}{\partial \boldsymbol{x}_p} = -(\nabla_{\boldsymbol{x}_p}\nabla_{\boldsymbol{w}}\mathcal{L}_{tr})(\nabla^2_{\boldsymbol{w}}\mathcal{L}_{tr})^{-1} \tag{3.8}$$

By inserting Eqs. 3.8 into 3.6, the gradient required to solve the bilevel optimization problem can be written as

$$\nabla_{\boldsymbol{x}_p}\mathcal{A} = \nabla_{\boldsymbol{x}_p}\mathcal{L} - (\nabla_{\boldsymbol{x}_p}\nabla_{\boldsymbol{w}}\mathcal{L}_{tr})(\nabla^2_{\boldsymbol{w}}\mathcal{L}_{tr})^{-1}\nabla_{\boldsymbol{w}}\mathcal{L} \tag{3.9}$$

Then, following a gradient ascent strategy, the update equation to compute the poisoning point is given by

$$\boldsymbol{x}_p = \Pi_\phi\left(\boldsymbol{x}_p + \eta\nabla_{\boldsymbol{x}_p}\mathcal{A}\right) \tag{3.10}$$

where η is the learning rate and Π_{ϕ} is a projection operator to map the current update of the poisoning point onto the feasible domain ϕ that models the constraints of the attacker in the outer optimization problem.

Synthetic Example

Figure 3.3 shows an example of the optimal poisoning attack strategy on a synthetic example with a logistic regression classifier. In this example, there are two classes, red and yellow dots, and the attacker aims to inject a poisoning point from the red class to maximize the overall classification error. To do so, the attacker's objective is evaluated in a separate (validation) dataset, $\mathcal{D}_{target}$, with data points drawn from the same distribution as the training points shown in the figure. Cross-entropy was used as the loss function for the attacker's objective, $\mathcal{A}$, and for the logistic regression classifier, $\mathcal{L}_{tr}$.

In Fig. 3.3a, the trajectory of the poisoning point (yellow line) is shown when applying the gradient ascent strategy. The background color map depicts the attacker's objective function $\mathcal{A}$. The black rectangle represents ϕ, the constraints imposed by the attacker to control the values of the features for the poisoning points. The red star represents the poisoning point injected by the attacker as a result of the bilevel optimization problem in Eq. 3.5. As expected, the poisoning point is located at the point within the feasible domain for which the cross-entropy is maximized. Additionally, the effect of the poisoning point on the decision boundary (solid red line) can be seen in comparison with the decision boundary before the attack (dashed red line).

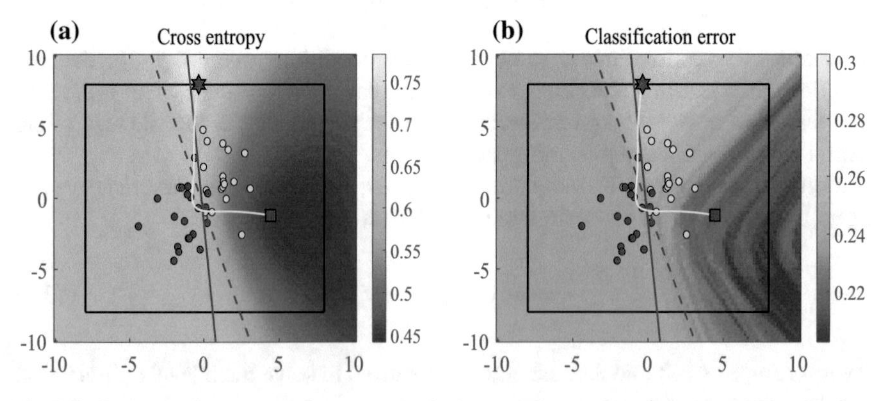

Fig. 3.3 Optimal poisoning attack on a synthetic dataset. The attacker aims to maximize the classification error by injecting one poisoning point (red). The black rectangle represents the attacker's constraints. The decision boundary before and after the attack is represented with dashed and solid red lines, respectively. The trajectory of the gradient ascent strategy followed to solve the optimization problem in Eq. 3.5 is depicted in yellow, and the final poisoning point is represented as a red star. The color map in (**a**) depicts the cross-entropy used to model the attacker's objective $\mathcal{A}$, evaluated in $\mathcal{D}_{target}$. The color map in (**b**) represents the classification error evaluated in the same dataset used for $\mathcal{A}$

Figure 3.3b shows the same scenario, but the color map is represented with the classification error evaluated in the validation dataset used by the attacker in $\mathcal{A}$. Attackers cannot maximize the classification error directly, because it is a non-differentiable function, hence the optimal attack strategy in Eq. 3.5 cannot be applied. Note, however, that the color map for the classification error is similar to the one depicted in Fig. 3.3a for cross-entropy. This shows that cross-entropy is good for maximizing the classification error.

In Fig. 3.4, an example of a poisoning attack is shown in MNIST,[8] a well-known benchmark for handwritten digit recognition in computer vision problems. To simplify the scenario, a binary classification problem with digits 1 and 7 is considered. Similar to the previous example, the attacker aims to maximize the classification error by maximizing the cross-entropy and targeting a logistic regression classifier. In this case, the attacker aims to inject a poisoning point labeled 7 to maximize the classification error. To do so, the attacker solves the bilevel optimization problem in Eq. 3.5 starting from an image representing 1 (see Fig. 3.4a). The resulting poisoning point, shown in Fig. 3.4b, have features related to both digits, 1 and 7, i.e., the shape of the poisoning point is designed in such a way that targets features that are important to distinguish the two digits.

Reducing Computational Complexity

Each update in Eq. 3.10 requires to solve the inner optimization problem to get the KKT conditions, i.e., the corresponding derivatives are evaluated for the values of $\boldsymbol{w}$ obtained after training the learning algorithm. On the other hand, computing and inverting the Hessian $\nabla^2_{\boldsymbol{w}}\mathcal{L}_{tr}$ scales in time as $\mathcal{O}(d^3)$ and in memory as $\mathcal{O}(d^2)$, being d the cardinality of $\boldsymbol{w}$. Moreover, the computation of Eq. 3.9 requires solving one linear system per feature in the poisoning point, which is also computationally expensive, and in some cases, infeasible.

Following a similar approach to [18], the computational complexity of Eq. 3.9 can be reduced by applying a conjugate gradient descent to solve a simpler linear system, which is obtained by a trivial reorganization of the terms in the second part of Eq. 3.9. This can be achieved by first solving the linear system given by

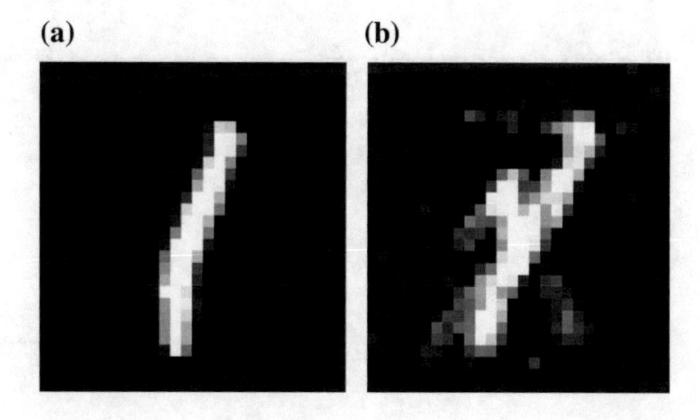

Fig. 3.4 Example of the optimal poisoning attack in MNIST dataset (using digits 1 and 7). **a** Initial point chosen by the attacker. **b** Final poisoning point after solving the bilevel optimization problem for an optimal attack

[8]http://yann.lecun.com/exdb/mnist/

$$(\nabla^2_{\boldsymbol{w}} \mathcal{L}_{tr})\boldsymbol{v} = \nabla_{\boldsymbol{w}} \mathcal{L} \tag{3.11}$$

and then compute the gradient as

$$\nabla_{\boldsymbol{x}_p} \mathcal{A} = \nabla_{\boldsymbol{x}_p} \mathcal{L} - (\nabla_{\boldsymbol{x}_p} \nabla_{\boldsymbol{w}} \mathcal{L}_{tr})\boldsymbol{v} \tag{3.12}$$

The computation of the matrices $\nabla_{\boldsymbol{x}_p} \nabla_{\boldsymbol{w}} \mathcal{L}_{tr}$ and $\nabla^2_{\boldsymbol{w}} \mathcal{L}_{tr}$ can be avoided by using the following Hessian vector products [19]:

$$\begin{aligned}(\nabla_{\boldsymbol{x}_p} \nabla_{\boldsymbol{w}} \mathcal{L}_{tr})z &= \lim_{h \to 0} \frac{1}{h} \big(\nabla_{\boldsymbol{x}_p} \mathcal{L}_{tr}(\boldsymbol{x}_p, \boldsymbol{w} + hz) - \nabla_{\boldsymbol{x}_p} \mathcal{L}_{tr}(\boldsymbol{x}_p, \boldsymbol{w})\big) \\ (\nabla^2_{\boldsymbol{w}} \mathcal{L}_{tr})z &= \lim_{h \to 0} \frac{1}{h} \big(\nabla_{\boldsymbol{w}} \mathcal{L}_{tr}(\boldsymbol{x}_p, \boldsymbol{w} + hz) - \nabla_{\boldsymbol{w}} \mathcal{L}_{tr}(\boldsymbol{x}_p, \boldsymbol{w})\big)\end{aligned} \tag{3.13}$$

Although these numerical tricks allow for a more efficient computation of the poisoning points, the inner optimization problem in Eq. 3.5 still has to be solved exactly for each iteration in the gradient ascent. This can be infeasible in large neural networks and deep learning systems, which require a lot of time to be trained. However, as proposed in [7], this limitation can be overcome by employing *back-gradient optimization*, a technique first proposed in the context of energy-based models [20], and later in deep learning architectures [35] for hyperparameter optimization, via solving bilevel optimization problems similar to the one in Eq. 3.5.

The underlying idea of this approach is to replace the inner optimization problem with a reduced set of iterations performed by the learning algorithm to update parameters $\boldsymbol{w}$. This requires updates to be smooth, similar to the case of gradient-based learning algorithms, such as neural networks and deep architectures. This way, it is possible to compute the gradients required in the outer optimization problem using $\boldsymbol{w}_T$, the parameters of the learning algorithm obtained from an incomplete optimization of the inner problem, truncated to T iterations [20].

In Algorithm 1, the standard gradient-descent approach is described, which is used to solve the incomplete inner optimization problem, limiting the number of iterations to T.

Algorithm 1: Gradient Descent

1 **Input:** initial parameters $\boldsymbol{w}_0$, learning rate η, $\mathcal{D}_{tr}$, $\mathcal{L}_{tr}$
  1: **for** $t = 0, \ldots, T - 1$ **do**
  2: $\quad \mathbf{g}_t = \nabla_{\boldsymbol{w}} \mathcal{L}_{tr}(\mathcal{D}_{tr}, \boldsymbol{w}_t)$
  3: $\quad \boldsymbol{w}_{t+1} \leftarrow \boldsymbol{w}_t - \eta\, \mathbf{g}_t$
  4: **end for**

Output: trained parameters $\boldsymbol{w}_T$

In Algorithm 2, the back-gradient optimization procedure is described that computes $\nabla_{x_p}\mathcal{A}$ with a recurrent algorithm similar to gradient descent with T iterations.

Algorithm 2: Back-Gradient Descent

Input: trained parameters $\boldsymbol{w}_T$, learning rate (inner problem) η, learning rate (outer problem) α, $\mathcal{D}_{tr}$, $\mathcal{D}_{target}$, poisoning point $(\boldsymbol{x}_p, y_p)$, attacker's function $\mathcal{L}$, learner's loss function $\mathcal{L}_{tr}$.
initialize $d\boldsymbol{x}_p \leftarrow \boldsymbol{0}$, $d\boldsymbol{w} \leftarrow \nabla_{\boldsymbol{w}}\mathcal{L}(\mathcal{D}_{target}, \boldsymbol{w}_T)$

1: **for** $t = T, \ldots, 1$ **do**
2: $d\boldsymbol{x}_p \leftarrow d\boldsymbol{x}_p - \eta\, d\boldsymbol{w}\nabla_{\boldsymbol{x}_p}\nabla_{\boldsymbol{w}}\mathcal{L}_{tr}(\boldsymbol{x}_p, \boldsymbol{w}_t)$
3: $d\boldsymbol{w} \leftarrow d\boldsymbol{w} - \eta\, d\boldsymbol{w}\nabla^2_{\boldsymbol{w}}\mathcal{L}_{tr}(\boldsymbol{x}_p, \boldsymbol{w}_t)$
4: $\mathbf{g}_{t-1} = \nabla_{\boldsymbol{w}_t}\mathcal{L}_{tr}(\boldsymbol{x}_p, \boldsymbol{w}_t)$
5: $\boldsymbol{w}_{t-1} = \boldsymbol{w}_t + \alpha\mathbf{g}_{t-1}$
6: **end for**

Output: $\nabla_{\boldsymbol{x}_p}\mathcal{A} = \nabla_{\boldsymbol{x}_c}\mathcal{L} + d\boldsymbol{x}_p$

This represents a significant computational improvement compared to traditional approaches, because it only requires to compute a reduced number of iterations for the learning algorithm. This allows poisoning attacks against large neural networks and deep learning systems [7].

3.3.3 Transferability of Poisoning Attacks

In adversarial machine learning, attack transferability is an important property, which can be defined as how effective an attack that targets a specific learning algorithm is when applied to a different algorithm. Thus, transferability of attacks between learning algorithms allows to analyze up to what extent the algorithms share the same vulnerabilities, which can be valuable when proposing defensive mechanisms that can be effective to mitigate the effect of attacks for a broader range of algorithms. Following the threat model described in Sect. 3.2, transferability attacks can be defined as partial knowledge attacks with surrogate models, i.e., attacks in which the attacker does not know the learning algorithm used by the defender.

Although transferability has been mainly analyzed in the context of evasion attacks [21] (see Sect. 3.4.2), experimental evidence of this property has also been shown in the context of poisoning attacks [7]. More precisely, it has been shown that poisoning attack strategies targeting specific learning algorithms are quite effective against other algorithms as well, particularly if they are similar. For example, attacks between different linear classifiers are quite transferable.

To illustrate this, Fig. 3.5 shows a synthetic example in a binary classification task, where the aim is to classify the red and the yellow dots with a linear classifier. The attacker attempts to inject a red poisoning point to maximize the error of the

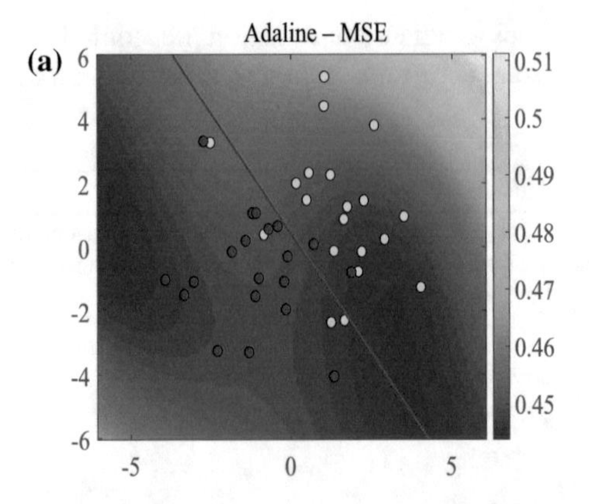

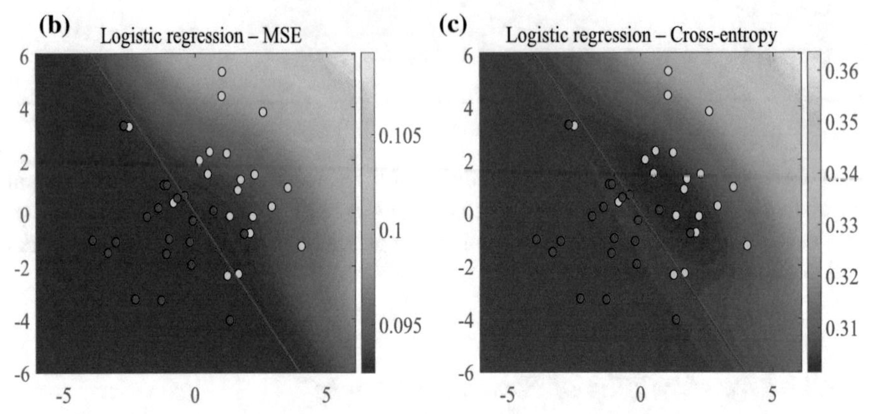

Fig. 3.5 Transferability of poisoning attacks. Synthetic example with a linear classification problem using three different configurations: **a** Adaline minimizing the MSE; **b** Logistic regression minimizing the MSE; **c** Logistic regression minimizing cross-entropy. In all the plots, the color map shows the attacker's loss $\mathcal{A}$ evaluated on a separate dataset $\mathcal{D}_{target}$. The red line represents the decision boundaries. Note that the maximum of the attacker's loss is located at the same position for the three plots, which means that the poisoning attacks are transferable between the three models

classifier.Thus, in Fig. 3.5a an *Adaline* classifier (sometimes also incorrectly referred as *Perceptron*) is used that minimizes the mean squared error (MSE)[9] over the set of training points. The color map shows the MSE for the attacker's loss function $\mathcal{A}$, which in this case is also the MSE evaluated over a separate dataset $\mathcal{D}_{target}$. The color map suggests that the maximum of this attacker's function is achieved for poisoning points in the upper-right corner of the plot.

Similarly, in Fig. 3.5b a similar color map is shown with the MSE evaluated on $\mathcal{D}_{target}$ when the defender is training a logistic regression classifier that aims to minimize the MSE over the same training set that was used by the previous Adaline

[9]Mean square error is the average of the squares of errors. It is a measure of estimator quality, is always non-negative, and the closer its value to zero the better.

classifier. Although the color map is slightly different, the maximum of the attacker's objective function is located at the same position as in the previous case. A similar result is obtained in Fig. 3.5c, in which both the attacker and the defender use cross-entropy as the loss function. Therefore, if the optimal poisoning attack strategy is applied in the three scenarios, the resulting poisoning attack point will be identical.

This example suggests that regardless of the loss function used by the attacker and the defender, and the type of linear classifier applied, the attack points are transferable. Therefore, even if the attacker does not have perfect knowledge about the learning algorithm used by the defender, the effect of the attack can still be significant providing that the surrogate model used by the attacker is reasonably similar.

Transferability between linear and non-linear classifiers has also been empirically shown in [7], in which attacks carried out against linear classifiers were able to effectively compromise (non-linear) neural network architectures.

3.3.4 Defense Against Poisoning Attacks

Defensive techniques against data poisoning have been less explored in the literature compared to attack strategies, and it remains unclear how to effectively defend against some of these attacks.

Reject On Negative Impact (RONI) defense [14] was one of the first algorithms proposed to mitigate the effect of data poisoning via detecting and discarding samples in the training dataset that have a negative impact on the classifier's performance. RONI requires the retraining of the algorithm using subsets of the training dataset, which, similar to Leave-One-Out validation procedures, uses all the training samples in the subset except one, which is considered for validation. Next, RONI measures the validation performance for each sample in each subset, and removes the points that have the most negative impact on the classifier. Although this technique has been shown to be effective against some types of poisoning attacks [14], it is computationally expensive, as it requires the learning algorithm to be retrained for each point in the training set (even if small subsets are used to speed up the computation). This might be infeasible for deep networks and for applications that require large training datasets. Additionally, RONI may suffer from *overfitting* if the training subsets used by the algorithm are small compared to the number of features. In such cases, the validation performance may not be a reliable indicator to detect malicious data points.

Outlier detection and adequate data prefiltering can also help to reduce the impact of poisoning attacks, especially in cases when the attacker's constraints are not properly defined. For example, although the experimental results of the optimal poisoning attacks in [7, 15–17] show that the attacks considerably degrade the performance of machine learning classifiers, they do not consider any detectability constraint; in other words, they only check whether the generated attack points are valid points. The lack of detectability constraints in optimal attack strategies can lead to poisoning points that are quite different from the genuine ones [12], which can be detected and removed from the training dataset by applying outlier detection.

Relying on this argument, in the context of generative models, the authors in [22] proposed a two-step defense strategy for logistic regression classifiers. First, outlier detection is applied, followed by the training of the algorithm to solve an optimization problem based on the correlation between the classifier and the labels. Although this approach is effective, its main shortcoming is that the defender is assumed to know a fraction of the poisoning examples in advance, and even though this is not available in practical applications, the performance of the algorithm is sensitive to this value.

An outlier detection scheme to defend against data poisoning in classification tasks is proposed in [12], with which it is possible to train outlier detectors for each class by relying on a small fraction of trusted data points. This strategy is effective to mitigate the effect of poisoning attacks in case the attacker does not model specific attack constraints, but its capabilities to detect more constrained attacks are limited. This is the case of label flipping attacks, where the attacker can only manipulate the labels of the poisoning points. In this case, the previous algorithm is only capable to detect and remove the poisoning points far from the genuine points of the corresponding class, which is not always the case. Additionally, this technique requires to curate a small fraction of the training points, i.e., to check that some of the training points are labeled adequately, and that they are not malicious. This assumption might be reasonable in many application domains, although it can be limiting in some contexts. Relying on outlier detection, the authors in [23] provided an upper bound on the performance of binary classifiers under data poisoning, also assuming that some data preprocessing is performed before training. However, the bound was computed using a surrogate loss function rather than using the classification error, so that the computed upper bound can be quite loose.

Other defense techniques have also been proposed against less aggressive strategies, such as label flipping attacks. In [24], a defensive mechanism based on *k-Nearest Neighbors* is proposed to relabel possibly malicious data points based on the labels of their neighboring samples in the training set. However, this can result in limited performance for attacks in which subsets of poisoning points are close. In [25], *influence functions* from robust statistics have been proposed both to generate poisoning examples and as a countermeasure against label flipping attacks. Base on these, it is possible to identify the most dangerous samples in the training set that would need manual inspection. Influence functions provide an estimate of the influence of each training sample on the performance of the machine learning system. In contrast to RONI, the computation of the influence functions do not require to retrain the algorithm for each training data point, only for some gradient calculations, thereby reducing the computational complexity significantly. In other words, influence functions can be applied to mitigate the effect of label flipping attacks in deep learning systems.

3.4 Attacks at Test Time

Evasion attacks are attacks produced at test time, in which the attacker aims to manipulate the input data to produce an error in the machine learning system. In contrast to data poisoning, evasion attacks do not alter the behavior of the system; instead, they exploit its blind spots and weaknesses to produce the desired errors. Similar to the case of poisoning, these attacks have been mainly analyzed in the context of machine learning classification, and, similar to the previous section, machine learning classifiers are considered here as well.

Evasion attacks can exploit the following two vulnerabilities in the machine learning system:

- By leveraging regions of the feature space not supported by the training data (by generating attack points that are quite different from the points used to train the learning algorithm). In these regions, the algorithm can produce quite unexpected predictions. However, these attacks can be easily mitigated by adequate data prefiltering or by outlier detection. Consequently, points that are considered outliers are rejected by the system. For this reason, this section does not analyze these scenarios.
- By exploiting the weaknesses of the learning algorithm (regions of the feature space for which the learned decision boundary differs from the true unknown decision boundary that separates the classes in an optimal way). This is because the number of data points used for training the algorithm is finite and/or the learning algorithm has limited capacity to solve the classification problem (for example when using a linear classifier to solve a non-linear classification problem or by using small neural networks with limited expressive power). In contrast, in noisy classification problems in which the classes cannot be perfectly separated, the attacker can use regions with more overlapping between classes to facilitate the evasion.

In Fig. 3.6, a synthetic example is shown in a binary classification task, which illustrates the aforementioned weaknesses. The blue curve represents the true unknown decision boundary that could be obtained by having an infinite number of training points and a learning algorithm with enough capacity to solve the classification problem. The red curve depicts the decision boundary learned by a machine learning classifier on the training points shown in the figure. In this case, the algorithm is capable of correctly classifying the training points, which means that the algorithm is quite competent to solve the classification task. However, when comparing this w.r.t. the true decision boundary, there are regions (shaded in gray) in which the learning algorithm differs (areas where the learning algorithm is incorrect). These *adversarial regions* can be exploited to carry out successful evasion attacks.

Even if the machine learning classifier has enough expressive power to solve the classification task, in some cases these adversarial regions produce scarce data, for example due to the probability density of the data being very low in these regions. Hence, the chance of having training data points in these areas is also very low. This problem gets more severe as the number of features used by the dataset increases:

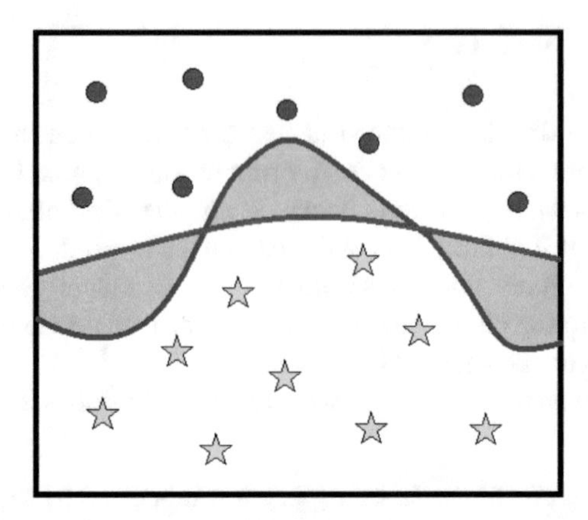

Fig. 3.6 Motivation of evasion attacks. In this binary classification problem, the true (unknown) decision boundary that optimally separates the two classes is depicted with a blue curve. Because the training dataset, given by the red dots and the yellow stars, is finite, the learned decision boundary is represented with a red curve. Although the learned classifier perfectly separates the training data, it differs from the true decision boundary. The gray areas represent the regions for which the machine learning classifier will make a mistake, which can be exploited by attackers to carry out successful evasion attacks

learning becomes more challenging, and a larger dataset is needed to have a dataset that actually represents the underlying data distribution.

Evasion attacks have recently started targeting deep neural networks, and it turned out that they are very vulnerable to this kind of threat. The associated incidents confirm the existence of *adversarial examples*, i.e., samples that can be misclassified by the deep learning system when adding imperceptible distortions to genuine images (which are correctly classified by the system). The existence of adversarial examples was first noticed in the context of image classification [9], where it was shown that undetectable modifications in an image can produce different predictions in deep learning classifiers. Then, [10, 26] provided a more comprehensive analysis of this threat, describing efficient evasion attack strategies capable of deceiving deep neural networks.

From a practical perspective, evasion attacks are considered relevant threats and can hinder the penetration of machine learning technologies to some application domains. For instance, if deep networks are used to perform image analysis in self-driving cars, attackers can exploit the vulnerabilities of the learning algorithms to produce undesirable actions, as for example by (physically) manipulating traffic signs on the road [27].

The rest of this section describes different evasion attack scenarios against machine learning classifiers according to the attacker's goal. It is also explained how to compute evasion attacks by solving a constrained optimization problem, showing some examples on MNIST. Similar to the case of data poisoning, the transferability properties of evasion attacks are also described. Finally, some mechanisms that can help to mitigate the effect of these attacks are described, although this area has challenges yet to be addressed.

3.4.1 Evasion Attack Scenarios

Using the threat model in Sect. 3.2 and following a treatment similar to what was used for poisoning attacks, evasion attack scenarios are described in multi-class classification systems according to the type of errors that the attacker wants to produce [28]. More specific attack scenarios can also be derived from these.

Error-Agnostic Evasion Attacks

In multi-class classification, the predicted class c^* is often considered to be the class whose discriminant function for a given input sample $\boldsymbol{x}$, $f_k(\boldsymbol{x})$ with $k = 1, \ldots, c$, is maximum:

$$c^* = F(\boldsymbol{x}) = \arg\max_{k=1,\ldots,c} f_k(\boldsymbol{x}) \tag{3.14}$$

In error-agnostic evasion attacks, also known as indiscriminate evasion, the attacker aims to produce a misclassification at test time, regardless of the class predicted by the classifier. For instance, in a face recognition system, an attacker may want to prevent the system from recognizing him, regardless of the incorrect identity predicted by the algorithm.

In this scenario, error-agnostic evasion attacks can be formulated as the following optimization problem:

$$\begin{aligned} \boldsymbol{x}_e^* \in \arg\min_{\boldsymbol{x}_e \in \phi(\boldsymbol{x}_e)} f_k(\boldsymbol{x}_e) - \max_{j \neq k} f_j(\boldsymbol{x}_e), \\ \text{s.t.} \;\; d(\boldsymbol{x}, \boldsymbol{x}_e) \leqslant d_{max} \end{aligned} \tag{3.15}$$

where $f_k(\boldsymbol{x}_e)$ denotes the discriminant function associated with the true class evaluated in $\boldsymbol{x}_e$, $\max_{j \neq k} f_j(\boldsymbol{x}_e)$ is the discriminant function from the set of incorrect classes whose value is higher. The constraint of the optimization problem, $d(\boldsymbol{x}, \boldsymbol{x}_e) \leqslant d_{max}$, limits the maximum perturbation, d_{max}, allowed between the original sample (the one the attacker wants to modify to evade the system), $\boldsymbol{x}$, and the attack point, $\boldsymbol{x}_e$, given a distance metric d. In the research literature, typically the ℓ_2, ℓ_1, and ℓ_∞ distances are used to constrain the perturbation of the adversarial examples. Additionally, attack point $\boldsymbol{x}_e$ is also restricted to the region of valid points determined by ϕ, which can also model attack scenarios in which the attacker can only modify some of the features used by the learning algorithm.

Error-Specific Evasion Attacks

In error-specific evasion attacks, the attacker aims to evade the system to produce a specific type of error. For example, in traffic sign detection for self-driving cars, the attacker may want a stop sign to be misclassified as a yield (give way) sign. These attacks are usually called *targeted evasion attacks* in the literature.

These attacks can be formulated as the optimization problem given by

$$\begin{aligned} \boldsymbol{x}_e^* \in \arg\min_{\boldsymbol{x}_e \in \phi(\boldsymbol{x}_e)} & \left(\max_{j \neq k_e} f_j(\boldsymbol{x}_e) \right) - f_{k_e}(\boldsymbol{x}_e), \\ \text{s.t.} \ \ & d(\boldsymbol{x}, \boldsymbol{x}_e) \leqslant d_{max} \end{aligned} \tag{3.16}$$

where $f_{k_e}(\boldsymbol{x}_e)$ is the discriminant function of the target class k_e, i.e., the class to which the adversarial example should be assigned. Thus, $\max_{j \neq k_e} f_j(\boldsymbol{x}_e)$ is the discriminant function, different from the target class, with the highest value. The constraints are defined the same way as the error-agnostic attacks formulated in Eq. 3.15.

Synthetic Example

Figure 3.7 shows three different attack scenarios on a synthetic example for a classification problem with three classes, using a multi-class logistic regression as the machine learning classifier. In Fig. 3.7a, an error-agnostic attack scenario can be observed, in which the attacker aims to produce a misclassification of the green dot in the centre of the circle delimited by the dashed black line, which determines the maximum perturbation allowed for the attacker. In this example, the easiest way to evade the classifier is to misclassify the point as red, because the decision boundary between the green and the red class is closer to the original point. The attack point is depicted in green with a red border. Note that attacks with smaller perturbation would also be able to evade the system.

In Fig. 3.7b, an error-specific evasion attack is shown, in which the attacker attempts to misclassify the green point as blue. In this case, even though the decision boundary between the green and the blue classes is larger than in the previous case, the error-specific attack is still successful. However, if the maximum level of permitted perturbation is decreased, as shown in Fig. 3.7c, only an error-agnostic attack would be successful, because the decision boundary between the green and the blue classes is beyond the maximum budget allowed for the attacker.

3.4.2 Computing Evasion Attacks

The optimization problems in Eqs. 3.15 and 3.16 can be solved using a gradient descent strategy. This involves the update equation to compute the evasion point as follows:

$$\boldsymbol{x}_e = \Pi_{\phi, d_{max}}(\boldsymbol{x}_e - \eta \nabla_{\boldsymbol{x}_e} \mathcal{A}) \tag{3.17}$$

where $\mathcal{A} = f_k(\boldsymbol{x}_e) - \max_{j \neq k} f_j(\boldsymbol{x}_e)$ or $\mathcal{A} = \left(\max_{j \neq k_e} f_j(\boldsymbol{x}_e)\right) - f_{k_e}(\boldsymbol{x}_e)$ for error-generic and error-specific attacks, respectively. Similar to the case of poisoning attacks, at each iteration the updated point is projected onto the feasible domain through the projection operator $\Pi_{\phi, d_{max}}$, which considers the constraints defined by

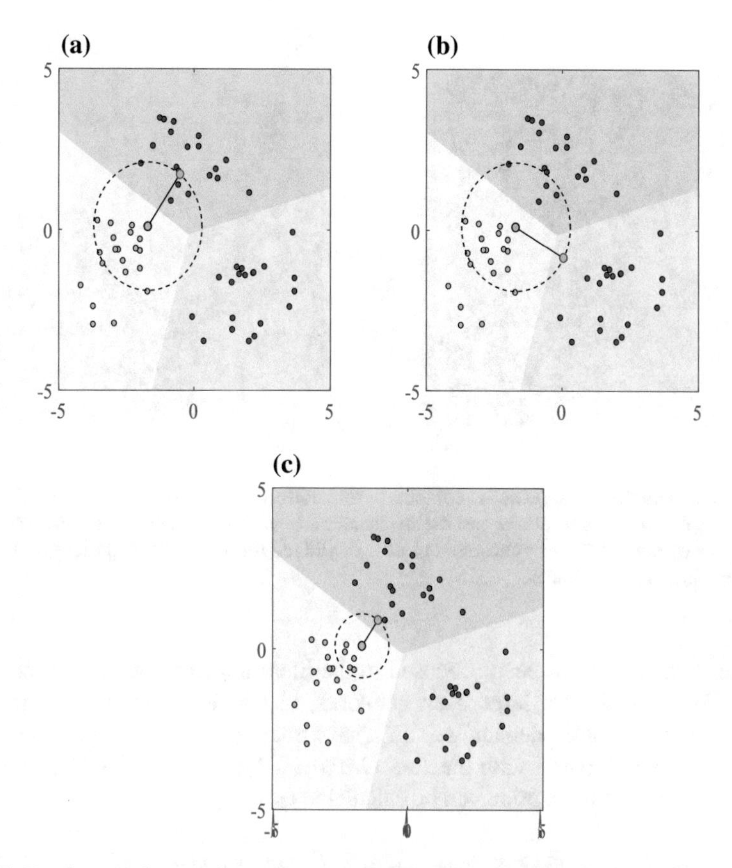

Fig. 3.7 Evasion attack scenarios: **a** Error-agnostic evasion: the green dot is modified in a way that it is classified as red (the dashed black line depicts the attacker's constraints). **b** Error-specific evasion: the attacker aims to misclassify the green dot as blue. **c** Evasion with a reduced budget: the selected green dot can only be misclassified as red

ϕ, i.e., the evasion point should be a valid point, and d_{max}, the maximum perturbation permitted according to the distance $d(\cdot, \cdot)$.

Figure 3.8 shows the trajectory of the evasion point when applying gradient descent on the synthetic scenarios in Fig. 3.7a and b for both error-agnostic and error-specific attacks. The color maps depict the cost ($\mathcal{A}$) the attacker tries to minimize. The blue regions correspond to lower $\mathcal{A}$ values. In the error-agnostic scenario in Fig. 3.8a, the blue regions encompass the areas close to the red and the blue points, i.e., both regions can be used to achieve evasion. However, given the initial position of the evasion point and the attacker's constraints, the gradient descent algorithm finds it easier to go to areas near the red points, because the decision boundary is closer. Figure 3.8b shows that for the error-specific attack, during which the attacker aims to classify the green dot as blue, only the region close to the blue points have low values for cost function $\mathcal{A}$.

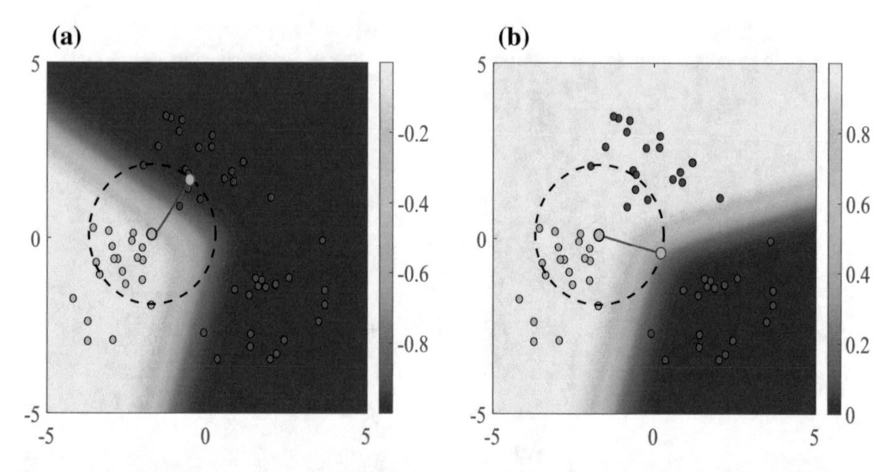

Fig. 3.8 Gradient descent approach to compute the evasion points in the scenario shown in Figs. 3.7a and b. The color map represents the cost the attacker aims to minimize for each scenario. The dashed black lines depict the attacker's constraints, and the solid red lines show the trajectory of the points when applying gradient descent

Other strategies have been proposed in the literature to compute evasion attack points. For example, for large deep networks, in which computing the gradients can be computationally demanding, an approximate (suboptimal) solution can be produced in one iteration with the *Fast Gradient Sign Method (FSGM)* [26], with which the evasion attack point can be calculated as

$$\boldsymbol{x}_e = \boldsymbol{x} + d_{max}\,\mathrm{sign}(\nabla_{\boldsymbol{x}}\mathcal{L}_{tr}((\boldsymbol{x}, y), \boldsymbol{w})) \tag{3.18}$$

where $\mathcal{L}_{tr}$ is the loss function minimized by the learning algorithm evaluated on the initial point $\boldsymbol{x}$ (with its initial label y). The formulation of this attack only considers ℓ_∞ scenarios in which each feature in the evasion point can be incremented or decremented by no more than d_{max}.

A different formulation of the optimization problem for error-specific attacks, Eq. 3.16, is proposed in [10], for which the objective is defined in terms of perturbation, and the score of the predicted evasion point is modeled as a constraint as follows:

$$\begin{aligned} \delta^* \in \arg\min{}_\delta ||\delta||_p \\ \text{s.t. } F(\mathbf{x} + \delta) = y^* \end{aligned} \tag{3.19}$$

where the attacker aims to minimize the perturbation δ (according to some norm p), subject to the fact that the evasion point $\boldsymbol{x}_e = \boldsymbol{x} + \delta$ should be classified as y^*, i.e., $F(\mathbf{x} + \delta) = y^*$. Additional constraints should also be considered to ensure that the evasion point is a valid point. This optimization problem was reformulated in [8] using Lagrangian multipliers to include the constraints in the optimization objective, and providing different scoring functions to model those constraints.

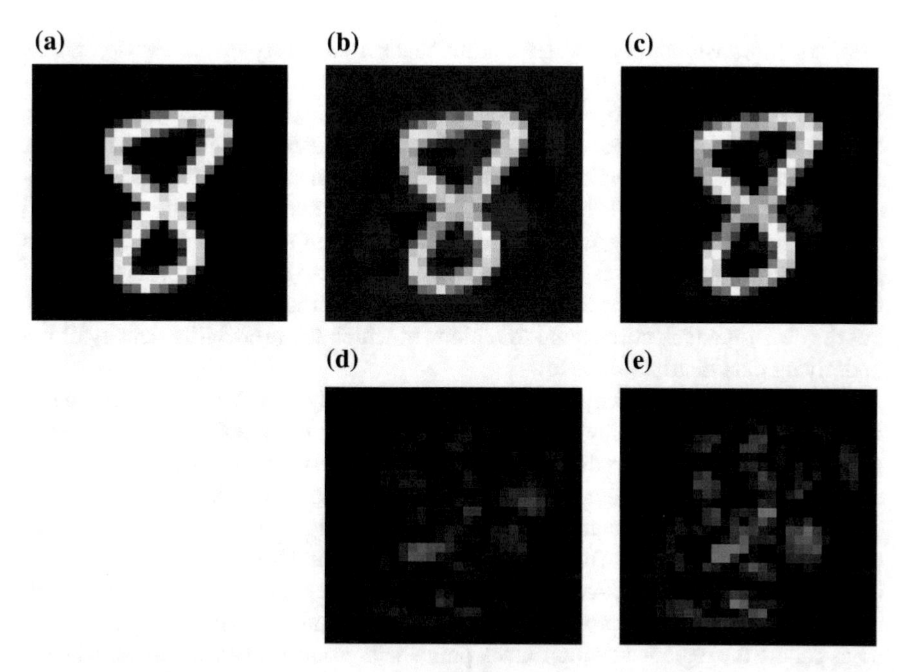

Fig. 3.9 Example of evasion attacks in the MNIST dataset. **a** Original image, correctly classified as an 8. **b** ℓ_2 norm evasion attack, limiting the ℓ_2 norm of the perturbation to 1.5. The digit is now classified as 6. **c** ℓ_∞ norm evasion attack, limiting the ℓ_∞ norm to 0.3. The digit is also classified as 6. **d** Perturbation introduced by the attacker in the ℓ_2 norm attack (the scale is multiplied by a factor of 2). **e** Perturbation introduced by the attacker in the ℓ_∞ norm attack (the scale is also multiplied by a factor of 2)

Example in the MNIST Dataset

Similar to the case of poisoning attacks in Sect. 3.3, Fig. 3.9 shows two examples of evasion points in the MNIST dataset for attacks using ℓ_2 and ℓ_∞ to model the attacker's constraints. To be more precise, Fig. 3.9a–c show the original point before the attack, and the resulting evasion points for the ℓ_2 and ℓ_∞ attacks, respectively. Before the attack, the image was correctly classified as 8, but after both attacks, the label of the corresponding images became 6. Figure 3.9d, e show the perturbations (augmented by a factor of 2) added to the original point in the attacks. Note that the digit represented in both Fig. 3.9b, c is 8, with an insignificant amount of background (adversarial) noise.

3.4.3 Transferability of Evasion Attacks

Section 3.3.3 introduced the concept of transferable attacks, which are attacks that can be used against multiple machine learning algorithms. In the context of evasion, attack transferability implies that an attacker can carry out an evasion attack on a surrogate model and transfer the attack points to the victim's model successfully with limited knowledge of the victim's algorithm. This enables the attacker to achieve successful black box attacks while reducing the number of queries required, thereby hindering the detection of the attack. It has been shown empirically that evasion attacks are transferable between different families and configuration of learning algorithms [21], especially for algorithms that share common characteristics. Attacks are also transferable in cases when the attacker uses surrogate datasets. As explained earlier, this is because some the regions that can be leveraged by the attacker to carry out the evasion attack correspond to regions in which the probability density of the underlying data distribution is low.

Figure 3.10 illustrates why evasion attacks are usually transferable between learning algorithms. Similar to the example in Fig. 3.6, for a binary classification problem, given a set of data points (red dots and yellow stars), two non-linear machine learning classifiers are trained on the same dataset. The corresponding decision boundaries are represented with green and red curves. Similar to Fig. 3.6, the blue curve represents the optimal (unknown) decision boundary that could be obtained for an infinite number of training points. As shown in Fig. 3.10, for the same training dataset, both non-linear classifiers may produce similar decision boundaries. Then, the gray shaded areas denote the regions in which attack points will produce an error with both learning algorithms. In contrast, the yellow shaded areas represent regions for which the attack points would be only effective against one of the classifiers. Comparing the two sets of regions suggests that in most cases, attack points generated for one of the classifiers will also be effective for the other one.

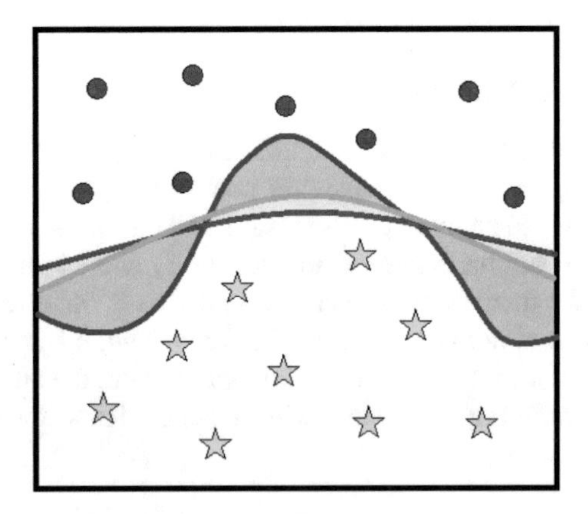

Fig. 3.10 Example of transferability in evasion attacks. The blue curve depicts the true (unknown) decision boundary. The red and green curves represent the decision boundaries learned by two non-linear machine learning classifiers trained on the red dots and yellow stars in the figure. The shaded gray areas show regions in which evasion attacks are effective against the two learning algorithms. Yellow regions depict attacks that are only effective against one of the classifiers. The comparison between the gray and yellow surfaces shows that most of the attacks committed against one learning algorithm can also compromise the other one effectively

3.4.4 Defense Against Evasion Attacks

Two main families of defensive techniques against evasion attacks have been investigated in the research community: the ones that aim to classify adversarial examples correctly, and the ones that attempt to detect adversarial examples. However, these defensive techniques are not effective to mitigate the effect of evasion attacks if the attacker targets a specific defense [31]. For some of these techniques, attackers just have to slightly increase the distortion of the evasion points to succeed in evading the system.

Adversarial retraining is perhaps one of the most effective ways to partially mitigate the effect of evasion attacks. The idea behind adversarial retraining is to introduce adversarial examples in the training dataset (using the correct labels). It was initially proposed in [9], but the cost of computing large sets of adversarial examples is prohibitive if standard gradient-based approaches are used, especially in deep learning systems. However, the computational complexity can be reduced by using the FGSM [26] in Eq. 3.18, which does not require an iterative search for the evasion point. Although this reduces the attack surface, adversarial retraining does not systematically eliminate all possible adversarial regions. Variants of adversarial retraining have also been proposed [29, 30], but they have proven to be inefficient across datasets [31].

Other defense techniques use *gradient masking* to hide the successful attacks when computing the gradients required to solve the optimization problems in Eqs. 3.15 and 3.16. Papernot et al. proposed *Defensive Distillation*, a training procedure that aims to smooth the model's decision surface in the adversarial directions that can be exploited by attackers [32]. However, as shown in Sect. 3.4.3, attack points are often transferable, meaning that attackers can generate adversarial examples using a surrogate model, and then transfer them to the victim's model with high probability of success. Additionally, it is shown in [33] that Defensive Distillation is no more robust against evasion attacks than unprotected neural networks.

Dimensionality reduction can also be applied to reduce the attack surface and mitigate the effect of evasion attacks. The authors in [34] propose to use *Principal Component Analysis (PCA)* for reducing the number of effective components used by the classifier. This means an inherent trade-off between the accuracy and the security of learning algorithms. You see, reducing the number of components can negatively affect the performance of the classifier, but a large number of components can provide more flexibility to the attacker to generate the adversarial examples. Experimental results in [34] show the effectiveness of dimensionality reduction to mitigate the effect of evasion attacks, although the authors in [31] state that this defense is not more secure than convolutional neural networks (CNNs). This can be argued, considering the experimental evaluation in [31], which compared the security of various machine learning algorithms (PCA applied to a standard deep neural network compared to a CNN), not proving that the defense proposed in [34] is not effective compared to algorithms in the same family.

A more comprehensive list and description of similar defense mechanisms and their limitations can be found in [31].

3.5 Conclusion

This chapter has shown vulnerabilities that can be exploited to compromise machine learning systems. A threat model has been introduced that can be applied to model different attack scenarios according to the attacker's capabilities, goal, knowledge, and strategy. There are two broad attack classifications by the attacker's capability to manipulate the learning algorithm: poisoning and evasion attacks. During poisoning attacks, the attacker can manipulate the behavior of the learning algorithm by injecting misleading data into the training set. In contrast, evasion attacks aim to produce errors in the system at test time, leveraging the limitations and blind spots of the learning algorithms.Various attack scenarios have been demonstrated for both attack types depending on the goal of the attacker. Specifically, error-specific attacks aim to produce specific types of errors in the learning algorithm, while error-agnostic attacks simply aim to cause errors in the system, no matter what the type.

Even though various defense mechanisms have been proposed to mitigate both poisoning and evasion attacks, they are still limited, making defense against these threats an open research problem, which is especially true for evasion attacks, for which detection and classification of adversarial examples are extremely difficult.

Beyond the research challenge of better understanding the attacks and defense mechanisms against machine learning systems, a particularly important open question is *how to systematically test the security of machine learning systems*. In this sense, traditional machine learning design methodologies rely on performance indicators (such as accuracy or error rates) that are measured on a separate validation (or test) dataset, which has not been used to train the learning algorithm. Then, *Occam's razor* principle is often applied for model selection: the simplest models with similar performance are preferred. This design methodology implies a trade-off between complexity and performance. However, when considering security aspects, i.e., the possible presence of an attacker targeting the learning algorithms, this design methodology is no longer a viable option.

In the case of poisoning attacks, worse-case analysis can be used to compare the robustness of the algorithms for different levels of data poisoning. Optimal attack strategies can help achieve this assessment, but currently they cannot be applied to families of algorithms for which gradients are not available, such as in the case of decision trees and random forests, as well as very large neural networks, for which the computation of the optimal poisoning points can be very expensive. For this reason, better attack strategies are needed, which can provide more capable models for the detectability constraints expected from skilled attackers who try to remain undetected.

Testing robustness against evasion attacks is even more challenging. Evaluating the performance of the algorithms on a test dataset drawn from the natural underlying data distribution does not provide any robustness indicator for evasion attacks, only an estimation of the upper bound of system performance. Therefore, measuring the robustness of machine learning systems against evasion attacks requires verification instead to make sure that the system will behave as expected for a broad range of scenarios.

References

1. Muñoz González L, Lupu EC (2018) The secret of machine learning. ITNow 60(1):38–39. https://doi.org/10.1093/itnow/bwy018
2. McDaniel P, Papernot N, Celik ZB (2016) Machine learning in adversarial settings. IEEE Secur Priv 14(3):68–72. https://doi.org/10.1109/MSP.2016.51
3. Huang L, Joseph AD, Nelson B, Rubinstein BI, Tygar J (2011) Adversarial machine learning. In: Chen Y, Cárdenas A.A, Greenstadt R, Rubinstein B (eds) Proceedings of the 4th ACM Workshop on Security and Artificial Intelligence. ACM, New York, pp 43–58. https://doi.org/10.1145/2046684.2046692
4. Barreno M, Nelson B, Joseph AD, Tygar J (2010) The security of machine learning. Mach Learn 81(2):121–148. https://doi.org/10.1007/s10994-010-5188-5
5. Barreno M, Nelson B, Sears R, Joseph AD, Tygar JD (2006) Can machine learning be secure? In: Lin, F-C, Lee, D-T, Lin B-S, Shieh S, Jajodia S (eds) Proceedings of the 2006 ACM Symposium on Information, Computer and Communications Security. ACM, New York, pp 16–25. https://doi.org/10.1145/1128817.1128824
6. Biggio B, Fumera G, Roli F (2014) Security evaluation of pattern classifiers under attack. IEEE T Knowl Data En 26(4):984–996. https://doi.org/10.1109/TKDE.2013.57
7. Muñoz-González L, Biggio B, Demontis A, Paudice A, Wongrassamee V, Lupu EC, Roli F (2017) Towards poisoning of deep learning algorithms with back-gradient optimization. In: Thuraisingham B, Biggio B, Freeman DM, Miller B, Sinha A (eds) Proceedings of the 10th ACM Workshop on Artificial Intelligence and Security, pp 27–38. https://doi.org/10.1145/3128572.3140451
8. Carlini N, Wagner D (2017) Towards evaluating the robustness of neural networks. In: 2017 IEEE Symposium on Security and Privacy. IEEE Computer Society, Los Alamitos, CA, USA, pp 39–57. https://doi.org/10.1109/SP.2017.49
9. Szegedy C, Zaremba W, Sutskever I, Bruna J, Erhan D, Goodfellow I, Fergus R (2013) Intriguing properties of neural networks. arXiv:1312.6199
10. Papernot N, McDaniel P, Jha S, Fredrikson M, Celik ZB, Swami A (2016) The limitations of deep learning in adversarial settings. In: Proceedings of the 2016 IEEE European Symposium on Security and Privacy. IEEE Computer Society, Los Alamitos, CA, USA, pp 372–387. https://doi.org/10.1109/EuroSP.2016.36
11. Papernot N, McDaniel P, Goodfellow I, Jha S, Celik ZB, Swami A (2017) Practical black-box attacks against machine learning. In: Karri R, Sinanoglu O, Sadeghi A-R, Yi X (eds) Proceedings of the 2017 ACM on Asia Conference on Computer and Communications Security. ACM, New York, pp 506–519. https://doi.org/10.1145/3052973.3053009
12. Paudice A, Muñoz-González L, György A, Lupu EC (2018) Detection of adversarial training examples in poisoning attacks through anomaly detection. arXiv:1802.03041
13. Joseph AD, Laskov P, Roli F, Tygar JD, Nelson B (eds.) (2013) Machine learning methods for computer security. Dagstuhl Manif 3(1):1–30. http://drops.dagstuhl.de/opus/volltexte/2013/4356/pdf/dagman-v003-i001-p001-12371.pdf

14. Nelson B, Barreno M, Chi FJ, Joseph AD, Rubinstein BI, Saini U, Sutton CA, Tygar JD, Xia K (2008) Exploiting machine learning to subvert your spam filter. In: Proceedings of the 1st Usenix Workshop on Large-Scale Exploits and Emergent Threats, article no. 7. USENIX Association, Berkeley, CA, USA. https://www.usenix.org/legacy/event/leet08/tech/full_papers/nelson/nelson.pdf
15. Biggio B, Nelson B, Laskov P (2012) Poisoning attacks against support vector machines. In: Langford J, Pineau J (eds) Proceedings of the 29th International Conference on Machine Learning, pp 1807–1814. arXiv:1206.6389
16. Mei S, Zhu X (2015) Using machine teaching to identify optimal training-set attacks on machine learners. In: Proceedings of the Twenty-Ninth AAAI Conference on Artificial Intelligence. AAAI Press, Palo Alto, CA, USA, pp 2871–2877. https://www.aaai.org/ocs/index.php/AAAI/AAAI15/paper/viewFile/9472/9954
17. Xiao H, Biggio B, Brown G, Fumera G, Eckert C, Roli F (2015) Is feature selection secure against training data poisoning? In: Bach F, Blei D (eds) Proceedings of the 32nd International Conference on Machine Learning, pp 1689–1698
18. Do CB, Foo CS, Ng AY (2007) Efficient multiple hyperparameter learning for log-linear models. In: Proceedings of the 20th International Conference on Neural Information Processing Systems. Curran Associates, Red Hook, NY, USA, pp 377–384
19. Pearlmutter BA (1994) Fast exact multiplication by the Hessian. Neural Comput 6(1):147–160. https://doi.org/10.1162/neco.1994.6.1.147
20. Domke J (2012) Generic methods for optimization-based modeling. In: Proceedings of the 15th International Conference on Artificial Intelligence and Statistics, pp 318–326. http://proceedings.mlr.press/v22/domke12/domke12.pdf
21. Papernot N, McDaniel, P, Goodfellow I (2016) Transferability in machine learning: from phenomena to black-box attacks using adversarial samples. arXiv:1605.07277
22. Feng J, Xu H, Mannor S, Yan S (2014) Robust logistic regression and classification. In: Ghahramani Z, Welling M, Cortes C, Lawrence ND, Weinberger KQ (eds) Proceedings of the 27th International Conference on Neural Information Processing Systems, vol 1. MIT Press, Cambridge, pp 253–261
23. Steinhardt J, Koh PWW, Liang PS (2017) Certified defenses for data poisoning attacks. In: Guyon I, Luxburg UV, Bengio S, Wallach H, Fergus R, Vishwanathan S, Garnett R (eds) Advances in neural information processing systems 30 (NIPS 2017). Curran Associates, Red Hook, NY, USA, pp 3520–3532. http://papers.nips.cc/paper/6943-certified-defenses-for-data-poisoning-attacks.pdf
24. Paudice A, Muñoz-González L, Lupu EC (2018) Label sanitization against label flipping poisoning attacks. arXiv:1803.00992
25. Koh PW, Liang P (2017) Understanding black-box predictions via influence functions. In: Proceedings of the 34th International Conference on Machine Learning, pp 1885–1894. arXiv:1703.04730v2
26. Goodfellow I, Shlens J, Szegedy C (2015) Explaining and harnessing adversarial examples. arXiv:1412.6572
27. Evtimov I, Eykholt K, Fernandes E, Kohno T, Li B, Prakash A, Rahmati A, Song D (2017) Robust physical-world attacks on deep learning models. arXiv:1707.08945
28. Melis M, Demontis A, Biggio B, Brown G, Fumera G, Roli F (2017) Is deep learning safe for robot vision? Adversarial examples against the iCub Humanoid. In: ICCV Workshop on Vision in Practice on Autonomous Robots, Venice, Italy, 23 Oct 2017. arXiv:1708.06939
29. Grosse K, Manoharan P, Papernot N, Backes M, McDaniel P (2017) On the statistical detection of adversarial examples. arXiv:1702.06280
30. Gong Z, Wang W, Ku WS (2017) Adversarial and clean data are not twins. arXiv:1704.04960
31. Carlini N, Wagner D (2017) Adversarial examples are not easily detected: bypassing ten detection methods. In: Proceedings of the 10th ACM Workshop on Artificial Intelligence and Security. ACM, New York, pp. 3–14. https://doi.org/10.1145/3128572.3140444
32. Papernot N, McDaniel P, Wu X, Jha S, Swami A (2016) Distillation as a defense to adversarial perturbations against deep neural networks. In: 2016 IEEE Symposium on Security and Privacy.

IEEE Computer Society, Los Alamitos, CA, USA, pp 582–597. https://doi.org/10.1109/SP.2016.41

33. Carlini N, Wagner D (2016) Defensive distillation is not robust to adversarial examples. arXiv:1607.04311
34. Bhagoji AN, Cullina D, Mittal P (2017) Dimensionality reduction as a defense against evasion attacks on machine learning classifiers. arXiv:1704.02654v2
35. Maclaurin D, Duvenaud D, Adams R (2015) Gradient-based hyperparameter optimization through reversible learning. In: Bach F, Blei D (eds) Proceedings of the 32nd International Conference on Machine Learning, pp 2113–2122

Chapter 4
Patch Before Exploited: An Approach to Identify Targeted Software Vulnerabilities

Mohammed Almukaynizi, Eric Nunes, Krishna Dharaiya, Manoj Senguttuvan, Jana Shakarian, and Paulo Shakarian

Abstract The number of software vulnerabilities discovered and publicly disclosed is increasing every year; however, only a small fraction of these vulnerabilities are exploited in real-world attacks. With limitations on time and skilled resources, organizations often look at ways to identify threatened vulnerabilities for patch prioritization. In this chapter, an exploit prediction model is presented, which predicts whether a vulnerability will likely be exploited. Our proposed model leverages data from a variety of online data sources (white hat community, vulnerability research community, and dark web/deep web (DW) websites) with vulnerability mentions. Compared to the standard scoring system (CVSS base score) and a benchmark model that leverages Twitter data in exploit prediction, our model outperforms the baseline models with an F1 measure of 0.40 on the minority class (266% improvement over CVSS base score) and also achieves high true positive rate and low false positive rate (90%, 13%, respectively), making it highly effective as an early predictor of exploits that could appear in the wild. A qualitative and a quantitative study are also conducted to investigate whether the likelihood of exploitation increases if a vulnerability is mentioned in each of the examined data sources. The proposed model is proven to be much more robust than adversarial examples—postings authored by adversaries in the attempt to induce the model to produce incorrect predictions. A discussion on the viability of the model is provided, showing cases where the classifier achieves high performance, and other cases where the classifier performs less efficiently.

4.1 Introduction

An increasing number of software vulnerabilities are discovered and publicly disclosed every year. A *vulnerability* is a weakness in a software product that can be exploited by an attacker to compromise the confidentiality, integrity, or availability of the system hosting that product and cause harm [1]. The National Institute

M. Almukaynizi (✉) · E. Nunes · K. Dharaiya · M. Senguttuvan · J. Shakarian · P. Shakarian
Arizona State University, Tempe, AZ, USA
e-mail: malmukay@asu.edu

© Springer Nature Switzerland AG 2019

L. F. Sikos (ed.), *AI in Cybersecurity*, Intelligent Systems Reference Library 151,
https://doi.org/10.1007/978-3-319-98842-9_4

of Standards and Technology (NIST)[1] maintains a comprehensive list of publicly disclosed vulnerabilities in its National Vulnerability Database (NVD).[2] The NVD also provides information regarding target software products (CPE),[3] severity rating (CVSS)[4] in terms of exploitability and impact, and the date a vulnerability was published.

In 2016 alone, at least 6,000 vulnerabilities were disclosed in the NVD. This has risen to over 14,500 vulnerabilities in 2017. Once the vulnerabilities are publicly disclosed, the likelihood of their exploitation increases [2]. With limited resources, organizations often look to prioritize which vulnerabilities to patch by assessing the impact they will have on the organization if exploited. An *exploit* is defined as a piece of code that modifies the functionality of a system using an existing vulnerability [3]. In this chapter, the exploits that have been used to target systems in real-world attacks are referred to as *real-world exploits*. In contrast, *Proof-of-Concept (PoC) exploits* are typically developed to verify the existence of a reported flaw in order to reserve a CVE-ID or to illustrate how a vulnerability can be exploited. PoCs generally require additional functionalities to be weaponized and be useful by malicious hackers. Although the chances of detecting real-world exploits of a particular vulnerability if a PoC is already present are high, the presence of a PoC does not imply that it has been used in the wild.

To be on the safe side, standard risk assessment systems, such as Common Vulnerability Scoring System (CVSS), Microsoft Exploitability Index,[5] and Adobe Priority Rating[6] report many vulnerabilities to be severe. The foregoing systems are broadly viewed as guidelines to supply vulnerability management teams with tools that help in patch prioritization. One commonality across those systems is that they rank vulnerabilities based on historical attack patterns that are relevant to the technical details of vulnerabilities that are evaluated, rather than what hacker communities discuss and circulate in the underground forums and marketplaces. This does little to alleviate the problem, because the majority of flagged vulnerabilities will not be attacked [4].

Furthermore, current methods of patch prioritization appear to fall short [4]. Verizon reported that over 99% of breaches are caused by exploits to known vulnerabilities. Cisco also reported that "the gap between the availability and the actual implementation of such patches is giving attackers an opportunity to launch exploits."[7] For some vulnerabilities, the time window to patch the system is very small. For instance, exploits targeting the Heartbleed[8] bug in the OpenSSL[9]

[1]https://www.nist.gov

[2]https://nvd.nist.gov

[3]https://nvd.nist.gov/cpe.cfm

[4]https://nvd.nist.gov/vuln-metrics/cvss

[5]https://technet.microsoft.com/en-us/security/cc998259.aspx

[6]https://helpx.adobe.com/security/severity-ratings.html

[7]https://www.cisco.com/c/dam/m/en_ca/never-better/assets/files/midyear-security-report-2016.pdf

[8]http://heartbleed.com

[9]https://www.openssl.org

cryptographic software library were detected in the wild 21 hours after the vulnerability was publicly disclosed [5]. Hence, organizations need to efficiently assess the likelihood that a vulnerability is going to be exploited in the wild, while keeping the false alarm rate low.

NIST provides the National Vulnerability Database (NVD), which contains a comprehensive list of vulnerabilities. Yet, only a small fraction (less than 3%) of these vulnerabilities are exploited in the wild [4, 6–9]—a result confirmed in this chapter. In addition, a previous work has found that the CVSS score provided by NIST is not an effective predictor of vulnerabilities being exploited [4]. It has previously been proposed that other methods such as the use of social media [8, 10], dark web markets [11–13], and certain white hat[10] websites like Contagio[11] would be suitable alternatives. However, these approaches have their limitations. For instance, methodical concerns on the use of social media for exploit prediction were recently raised [14]; and data feeds for exploit and malware were limited to single sites and were only used for analysis to provide insights on economic factors of those sites [12, 13]. While other studies demonstrate the viability of data collection, they do not quantify the results of prediction [10, 11].

In this chapter, the potential of identifying software vulnerabilities that will likely be exploited in real-world cyberattacks is investigated. In this effort, cyberthreat intelligence feeds are used, which are aggregated from a variety of sources. This problem is directly related to patch prioritization. Previous works on online vulnerability mentions either studied the correlation between such feeds and the existence of exploits in the wild [4, 15], or developed machine learning models to predict the availability of Proof-of-Concept exploits [6, 14, 16]. However, only a small fraction of vulnerabilities having PoCs are actually exploited in the wild.

After reviewing the literature, including studies on data gathered from dark web and deep web [4, 11, 17–19], conducting analyses on data feeds collected from various online sources (e.g., SecurityFocus,[12] Talos),[13] and after over one hundred interviews with professionals working for managed security service providers (MSSPs),[14] firms specializing in cyber-risk assessment, and security specialists working for managed (IT) service providers (MSPs), three data sources have been identified that can represent the current threat intelligence used for vulnerability prioritization: (1) Exploit-DB (EDB)[15] contains information on Proof-of-Concept exploits for vulnerabilities provided by security researchers from various blogs and security reports, (2)

[10]Ethical (white hat) hacker is a person who practices hacking activities against some computer network to identify its weaknesses and assess its security, rather than having malicious intent or seeking personal gain.

[11]http://contagiodump.blogspot.com

[12]http://www.securityfocus.com

[13]https://www.talosintelligence.com/vulnerability_reports

[14]An MSSP is a service provider that provides its clients with tools that continuously monitor and manage wide range of cybersecurity-related activities and operations, which may include threat intelligence, virus and spam blocking, and vulnerability and risk assessment.

[15]https://www.exploit-db.com

Zero Day Initiative (ZDI)[16] is curated by the commercial firm TippingPoint and uses a variety of reported sources focusing on disclosures by various software vendors and their security researchers, and (3) a collection of information scraped from over 120 sites on DW sites from a system introduced in [20, 21] and currently maintained by the cybersecurity firm CYR3CON.[17] The intuition behind each of these feeds was not only to utilize information that was aggregated over numerous related sources, but also to represent feeds commonly used by cybersecurity professionals.

This chapter focuses on vulnerabilities that have publicly been disclosed in 2015–2016. The presented models employ supervised machine learning techniques using Symantec[18] attack signatures as ground truth.

- The utility of the proposed machine learning models in predicting exploits in the wild is demonstrated with true positive rate (TPR)[19] of 90% while maintaining a false positive rate (FPR)[20] of less than 15%. In addition, the proposed model is compared to a recent benchmark model that utilized online mentions for exploit prediction [8]. The proposed model achieves a significantly higher precision while maintaining a recall under the assumptions made. The performance of variants of the presented model is examined, when both temporal mixing and the case when only a single source is used are considered. The robustness of the presented model against various adversarial data manipulation strategies is also discussed.
- Using vulnerability mentions on EDB, ZDI, and DW, the increase in the vulnerability exploitation likelihood over vulnerabilities only disclosed on the NVD is studied. Results are provided to demonstrate the likelihood of exploitation given the vulnerability mention on EDB (9%), ZDI (12%) and DW (14%) and is compared to the NVD (2.4%). The availability of such information relative to the time an exploit is found in the wild is also studied.
- Exploited vulnerabilities are analyzed based on various other features derived from these data sources, such as the language used. Apparently, Russian language sites on the dark web discussing vulnerabilities are 19 times more likely to be exploited than random vulnerabilities, more likely than vulnerabilities described on websites in any of the other languages. Additionally, the probability of exploitation is investigated in terms of both data source and software vendor.

The rest of the chapter is organized as follows. State-of-the-art approaches are discussed in Sect. 4.2. Section 4.3 outlines some challenges related to the problem addressed in this chapter. Section 4.4 provides an overview of the presented exploit prediction model and describes the data sources used. Vulnerability analysis is discussed in Sect. 4.5. In Sect. 4.6, experimental results are provided for predicting the

[16]http://www.zerodayinitiative.com

[17]https://www.cyr3con.ai

[18]https://www.symantec.com

[19]TPR is a metric that measures the proportion of exploited vulnerabilities that are correctly predicted from all exploited vulnerabilities.

[20]FPR is a metric that measures the proportion of non-exploited vulnerabilities that are incorrectly predicted as being exploited from the total number of all non-exploited vulnerabilities.

likelihood of vulnerability exploitation. The robustness of the presented machine learning model in comparison with adversarial data manipulation is demonstrated in Sect. 4.7. A discussion on the viability of the proposed model and the cost of misclassification are provided in Sect. 4.8.

4.2 Related Work

Predicting cybersecurity events is one of those domains that have recently received a growing attention [6, 16, 22–24]. However, only little work in this line of research has been proposed so far compared to works proposed for detecting cyberthreats after they occur. Predicting whether a disclosed vulnerability will be exploited in the wild is important to organizations for prioritizing vulnerability patching. Previous studies attempted to address this problem using both a standard scoring system, in particular the Common Vulnerability Scoring System (CVSS), and machine learning techniques. An approach to predict the likelihood that a software contains a yet-to-be-discovered vulnerability was proposed by Zhang et al. [25]. In this study, the National Vulnerability Database (NVD) was leveraged to predict the time a next vulnerability will be discovered in a software. The results showed a poor predictive capability of the NVD data, because the approach only uses the Common Vulnerability Scoring System (CVSS) and the Common Platform Enumeration (CPE) as features, but not the NVD description. CVSS version 2.0 is a poor indicator of predicting whether a vulnerability will be exploited [4]. Apparently, deciding to patch a vulnerability because of a high CVSS score is equivalent to randomly guessing which vulnerability to patch. Moreover, integrating information about whether a PoC exploit is available is a good predictor of exploitation. In our approach, CVSS scores are used as indicators.

Our approach closely resembles previous work on using publicly disclosed vulnerability information as features to train machine learning models to predict whether a given vulnerability will be exploited. Bozorgi et al. [16] proposed a model that engineered features from two online sources, namely the now-discontinued Open Source Vulnerability Database (OSVDB) and the NVD to predict whether PoCs will be developed for a particular vulnerability. In their data, 73% of the vulnerabilities were reported as exploited, which is significantly higher than the ones reported in the literature (1.3%) [4, 7, 8]. The reason behind this is that the authors considered PoC as an indicator of whether the vulnerability will be exploited, which is not true in most cases. Using this assumption, and using a support vector machine classifier, their approach predicts whether vulnerabilities will have PoCs available, a problem that is different from the one studied in this chapter. A similar technique used the NVD as the data source, and Exploit-DB as the ground truth, with 27% vulnerabilities marked as exploited (have PoC exploits) [6]. They achieve high accuracy on a balanced dataset. Our analysis aims to predict vulnerabilities that will be used in real-world attacks and not just the ones that have PoC exploits available.

Building on the work on using publicly disclosed vulnerabilities, Sabottle et al. looked at predicting the exploitability based on vulnerabilities disclosed in Twitter[21] [8]. They collected Twitter data that have references to CVE-IDs and use these tweets[22] to compute features. A linear support vector machine (SVM) classifier was trained for prediction. As the source of ground truth data, Symantec threat signatures were used to label positive samples. Compared to previous predictive studies, even though [8] maintained the class ratio of 1.3% vulnerabilities exploited, this study used a resampled and balanced dataset to report the results. Additionally, the temporal aspect (training data should precede testing) of the tweets was not maintained while performing the experiments. This temporal intermixing caused future events to influence the prediction of past events, a practice that is known to lead to unrealistic predictions [14]. In this chapter, the class imbalance and the temporal aspect are respected while reporting the results.

In the literature, the impact of adversarial interference for hackers aiming to poison and evade the machine learning models has been discussed. Hao et al. proposed a prediction model to predict the web domain abuse based on features derived from the behavior of users at the time of registration [26]. They studied different evading strategies attackers can use, and demonstrated that evading attempts are expensive to attackers, and their model's reliance on different sets of features allows for limiting the decrease in false positive rate. Other researchers have also discussed the robustness of their models against adversarial attacks (e.g., [8, 22]). In this work, simulation experiments have been executed, which pointed out that the impact is very limited, as discussed in Sect. 4.7.

4.3 Preliminaries

To see how our proposed model works, some machine learning approaches and known challenges have to be discussed first.

4.3.1 Supervised Learning Approaches

The machine learning approaches used in this study are the following:

- **Support Vector Machine (SVM)**. Introduced by Cortes and Vapnik, SVM attempts to find a separating margin that maximizes the geometric distance between classes [27] (in the case of vulnerabilities, exploited and not exploited). The separating margin is called a hyperplane. When a separating plane cannot be found to distinguish between the two classes, the SVM cost function includes a regulariza-

[21] https://twitter.com

[22] Twitter posts, called *tweets*, are limited to 280 characters.

tion penalty and a slack variable for the misclassified samples. By varying these parameters, different trade-offs can be achieved between precision and recall.

- **Naïve Bayes Classifier (NB)**. Naïve Bayes is a probabilistic classifier, which uses Bayes theorem with independent attribute assumption. During training, the conditional probabilities of a sample of a given class having a certain attribute are computed. The prior probabilities for each class are also computed, i.e., the fraction of the training data belonging to each class. Naïve Bayes assumes that the attributes are statistically independent; therefore the likelihood for a sample S represented with a set of attributes a associated with a class c is given as $\Pr(c|S) = P(c) \times \prod_{i=1}^{d} \Pr(a_i|c)$.
- **Bayesian Network (BN)**. BN is a generalized form of NB such that not all features are assumed to be independent. Instead, variable dependencies are modeled in a graph leaned from the training data.
- **Decision Tree (DT)**. Decision tree is a hierarchical recursive partitioning algorithm. A decision tree is built by finding the best split attribute, i.e., the attribute that maximizes the information gain at each split of a node. In order to avoid overfitting, the terminating criteria is set to less than 5% of the total samples.
- **Logistic Regression (LOG-REG)**. Logistic regression classifies samples by computing the odds ratio. The odds ratio gives the strength of association between the attributes and the class.

4.3.2 Challenges of Exploit Prediction

The methodological issues of exploit prediction [14], along with finding a balance between conducting an evaluation under real-world conditions and with an adequate sample size, pose three challenges:

- **Class Imbalance**. As mentioned earlier, evidence of real-world exploits is found for only around 2.4% of the reported vulnerabilities. This skews the distribution towards one class in the prediction problem (i.e., *not exploited*). In such cases, standard machine learning approaches favor the majority class, leading to poor performance on the minority class. Some of the prior work in predicting the likelihood of exploitation considers the existence of PoCs as an indicator of real-world exploitations, which substantially increases the number of exploited vulnerabilities in the studies adopting this assumption [6, 14, 16]. However, out of the PoC exploits that are identified, only a small fraction are ever used in real-world attacks [7]—a result confirmed in this chapter (e.g., only about 4.5% of the vulnerabilities having PoCs were subsequently exploited in the wild). Other prior work uses class balancing techniques on both training and testing datasets and reports performance achieved using metrics like TPR, FPR, and accuracy[23]

[23]Note that these metrics are sensitive to the underlying class distribution and sensitive to the ratio of class rebalancing.

[22, 23]. Resampling the data to balance both classes in the dataset leads to training the classifier on a data distribution that is highly different than the underlying distribution. The impact of this manipulation, whether positive or negative, cannot be observed when testing the same classifier on a manipulated dataset, e.g., a testing set with the same rebalancing ratio. Hence, the prediction results of the proposed models in deployed, real-world settings are debatable. To confirm the impact of the highly imbalanced dataset used on the machine learning models, oversampling techniques (in particular SMOTE [28]) are examined. Note that the testing dataset is not manipulated because of the aim to observe a performance that can be reproduced in the settings of a model running on real-world deployment (e.g., streaming predictions). Doing so, only a marginal improvement is observed for some classifiers, as reported in Sect. 4.6.2, while other classifiers have shown a slightly negative impact when they are trained on an oversampled dataset.

- **Evaluating Models on Temporal Data**. Machine learning models are evaluated by training the model on one set of data, and then testing the model on another set that is assumed to be drawn from the same distribution. The data split can be done randomly or in a stratified manner, where the class ratio is maintained in both training and testing. Exploit prediction is a time-dependent prediction problem. Hence, splitting the data randomly violates the temporal aspect of the data—as events that happen in the future will now be used to predict events that happened in the past. Prior research efforts ignored this aspect while designing experiments [8]. In this work, this temporal mixing is avoided in most experiments. However, experiments with a very small sample size, in which this is not controlled, are included (this is because one of the used ground truth sources does not have date/time information). It is explicitly noted when this is the case.
- **Limitations of Ground Truth**. As mentioned, attack signatures reported by Symantec are used as the ground truth of the exploited vulnerabilities, similar to previous works [4, 8]. This ground truth is not comprehensive because the distribution of the exploited vulnerabilities over software vendors is found to differ from that for overall vulnerabilities (i.e., vulnerabilities that affect Microsoft products have a good coverage compared to products of other OS vendors). Although this source is limited in terms of coverage [8], it is still the most reliable source of exploited vulnerabilities because it reports attack signatures of exploits detected in the wild for known vulnerabilities. Other sources either report whether a software is malicious without proper mapping to the exploited CVE-ID (e.g., VirusTotal),[24] or rely on online blogs and social media sites to identify exploited vulnerabilities (e.g., SecurityFocus).[25] In this chapter, Symantec data is used while taking into account the false negatives. To avoid overfitting the machine learning model on this not-so-representative ground truth, the software vendor is omitted from the set of examined features.

[24] https://www.virustotal.com

[25] https://www.securityfocus.com There are many examples where attack signatures are reported by Symantec, but not reported by SecurityFocus. Also, there are vulnerabilities SecurityFocus reports as exploited, and those exist in software whose vendors are well-covered by Symantec, yet Symantec does not report them.

4.4 Exploit Prediction Model

Using machine learning models to address exploit prediction has interesting security implications in terms of prioritizing which vulnerabilities need to be patched first to minimize the risk of cyberattacks. Figure 4.1 gives an overview of the proposed exploit prediction model.

This model consists of the following phases:

1. **Data Collection**: Three data sources are used in addition to the NVD. These data sources are EDB (Exploit-DB), ZDI (Zero Day Initiative) and data mined from DW markets and forums, focusing on malicious hacking. Ground truth is assigned to the binary classification problem studied in this chapter using Symantec attack signatures of exploits detected in the wild. The data sources are described in Sect. 4.4.1.
2. **Feature Selection**: Features are extracted from each of the data sources. The features include bag-of-words features for vulnerability description and discussions on the DW, binary features that check for the presence of PoC exploits in EDB, vulnerability disclosures in ZDI and DW. Additional features are also included from the NVD, including CVSS scores and CVSS vectors.
3. **Prediction**: Binary classification is performed on the selected features to determine whether the vulnerability will likely be exploited or not. To address this classification problem, several standard supervised machine learning approaches are evaluated.

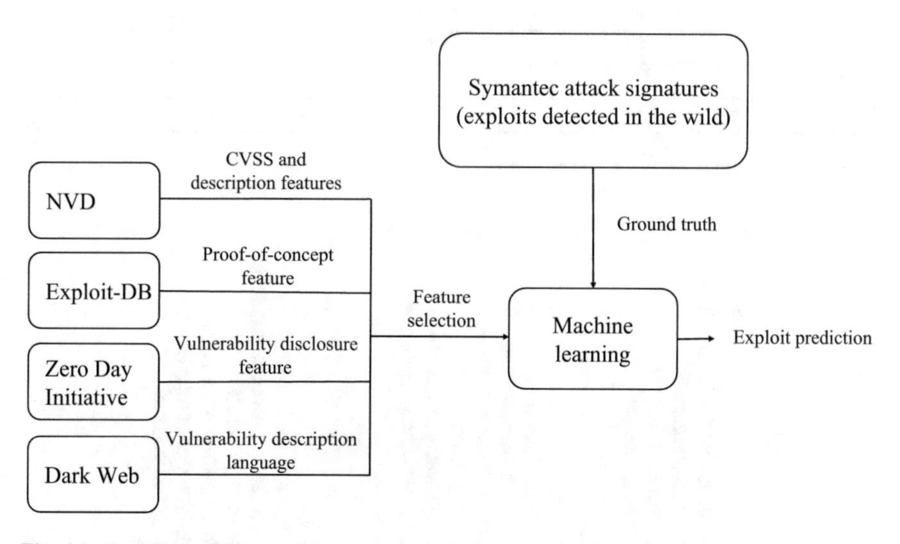

Fig. 4.1 Exploit prediction model

4.4.1 Data Sources

Our analysis combines vulnerability and exploit information from multiple open source databases, namely NVD, EDB, ZDI, and DW. The DW database was obtained from an API maintained by CYR3CON. Our experiments cover vulnerabilities published in 2015–2016. Table 4.1 shows the vulnerabilities identified from each of the data sources between 2015 and 2016, as well as the number of vulnerabilities that were exploited in real-world attacks.

A brief overview of each of these data sources is provided in the following sections.

4.4.1.1 National Vulnerability Database

The National Vulnerability Database is a database of publicly disclosed vulnerabilities. Each vulnerability is identified using a unique CVE-ID. Our dataset contains 12,598 vulnerabilities. Figure 4.2 shows the disclosure of vulnerabilities per month.

At the time of data collection in December 2016, there were only 30 vulnerabilities disclosed, hence the small bar at the end of 2016. For each vulnerability, the description, the CVSS score and vector, and the publication date were collected. Organizations often use the CVSS score to prioritize the vulnerabilities to patch. The

Table 4.1 Number of vulnerabilities (2015–2016)

Database	Vulnerabilities	Exploited	% exploited
NVD	12,598	306	2.4
EDB	799	74	9.2
ZDI	824	95	11.5
DW	378	52	13.8

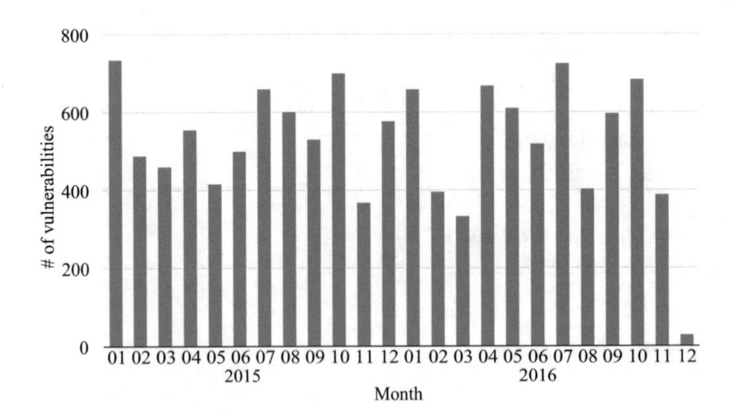

Fig. 4.2 Vulnerabilities disclosed per month

CVSS vector lists the components from which the score is computed. More details about CVSS components are provided in Sect. 4.4.2.

4.4.1.2 Exploit Database

The Exploit Database of the white hat community is an archive of PoC exploits maintained by Offensive Security.[26] PoC exploits for known vulnerabilities are reported with CVE-IDs of target vulnerabilities. Using the unique CVE-IDs from the NVD for the time period between 2015 and 2016, EDB was queried to find out whether a PoC exploit is available. The availability date of PoCs has also been recorded. By querying EDB, verified PoCs have been found for 799 of the vulnerabilities studied.

4.4.1.3 Zero Day Initiative

The Zero Day Initiative maintains a database of vulnerabilities that are identified and reported by security researchers. Reported software flaws are first verified by ZDI before disclosure. Monetary incentives are provided to researchers who report valid vulnerabilities. Before ZDI publicly discloses a vulnerability, the software vendors of the target products are notified and allowed time to implement patches. The ZDI database has been queried for the examined vulnerabilities, and found 824 common CVE-IDs between the NVD and ZDI. The publication date has also been noted.

4.4.1.4 Dark Web and Deep Web

The data collection infrastructure maintained by CYR3CON was originally described in [20]. In this paper, the authors built a system that crawls sites on DW, both marketplaces and forums, to collect data relating to malicious hacking. They first identify sites before developing scripts for automatic data collection. A site is put forward to script development after it has been determined whether the content is of interest (i.e., hacking-related) and is relatively stable. The population size of a site is observed, though not much decisive power is assigned to it. While a large population is an indicator for the age and stability of a site, a small population number can be associated with higher-value information (i.e., closed forums).

DW users advertise and sell their products on marketplaces. Hence, DW marketplaces provide a new avenue to gather information about vulnerabilities and exploits. Forums, on the other hand, feature discussions on kits of newly discovered vulnerabilities and exploits. Data related to malicious hacking is filtered from the noise and added to a database using a machine learning approach with high precision and recall. Not all exploits or vulnerability items in the database have an associated CVE number. First, the database was queried to extract all items with CVE mentions. Some

[26]https://www.offensive-security.com

vulnerabilities are mentioned in DW by their Microsoft Security Bulletin Number[27] (e.g., MS16-006). Every bulletin number was mapped to its corresponding CVE-ID, making ground truth assignment easy. These items can be both products sold on markets and posts extracted from forums discussing topics related to malicious hacking. 378 unique CVE mentions have been found between 2015 and 2016 on more than 120 DW websites, much more than what a previous work discovered [4]. In addition, the posting date and descriptions associated with all the CVE mentions have also been queried, including product title and description, vendor information, the entire discussion with the CVE mention, post author, and the topic of the discussion.

4.4.1.5 Attack Signatures (Ground Truth)

For ground truth, vulnerabilities that were exploited in the wild using Symantec antivirus attack signatures[28] and Intrusion Detection Systems (IDS) attack signatures[29] have been identified. The attack signatures are associated with the CVE-ID of the vulnerability that was exploited. These CVEs have been mapped to the CVEs mined from the NVD, EDB, ZDI, and DW. This ground truth indicates actual exploits that were used in the wild (not PoC exploits). For the NVD, around 2.4% of the disclosed vulnerabilities were exploited, which is consistent with the literature. In addition, for EDB, ZDI, and DW there is a significant increase in exploited vulnerabilities to 9%, 12%, and 14%, respectively. This shows that it is more likely to identify a vulnerability that will be exploited in the future if it has a PoC available in EDB or mentions in ZDI or DW. For this research, no data is available about the volume and frequency of the attacks carried by exploits; hence all exploited vulnerabilities are considered equally important. This assumption has been adopted by previous works as well [4, 8]. Additionally, the exploitation date of a vulnerability is defined as the date it was first detected in the wild. Symantec IDS attack signatures are reported without recoding the dates when they were first detected, but its anti-virus attack signatures are reported with their exploitation date. Between 2015 and 2016, 112 attack signatures have been reported without and 194 with their discovery date.

4.4.2 *Feature Description*

In this section, features from all the data sources discussed in the previous section are combined, including bag-of-words, numerical, categorical, and binary features (see Table 4.2).

These features are discussed in the following sections in more detail.

---

[27] https://technet.microsoft.com/en-us/security/bulletins.aspx

[28] https://www.symantec.com/security-center/a-z

[29] https://www.symantec.com/security_response/attacksignatures/

Table 4.2 Summary of features

Feature	Type
NVD and DW description	TF-IDF on bag of words
CVSS	Numeric and categorical
DW language	Categorical
Presence of PoC	Binary
Vulnerability mention on ZDI	Binary
Vulnerability mention on DW	Binary

4.4.2.1 NVD and DW Description

The NVD description provides information on the vulnerability and what it allows hackers to do when they exploit it. DW description often provides rich context on what the discussion is about, and is often synthesized from forums rather than marketplaces since items are described in fewer words. Patterns can be learned based on this textual content. The description of published vulnerabilities have been obtained from the NVD, and the DW database was queried for CVE mentions between 2015 and 2016. This description was appended to the NVD description with the corresponding CVE. Note that some of the descriptions on DW are in a foreign language as discussed in Sect. 4.5, which have been translated to English using the Google Translate API.[30] Then the text features were vectorized using the *frequency-inverse document frequency* (TF-IDF) model learned from the training set and used it to vectorize the testing set. TF-IDF creates a vocabulary of all the words in the description. The importance of a word feature increases with the number of times it occurs, but is normalized by the total number of words in the description. This eliminates common words from being important features. Our TF-IDF model was limited to the 1,000 most frequent words (using more word features has no benefit in terms of performance, however, it would increase the computational cost).

4.4.2.2 CVSS

NVD provides a CVSS score and the CVSS vector from which the score can be computed, indicating the severity of each disclosed vulnerability. CVSS version 2.0 was used, rather than version 3.0, because the latter is only present for a fraction of the studied vulnerabilities. The CVSS vector lists the components from which the score is computed. The components of the vector include Access Complexity, Authentication, Confidentiality, Integrity, and Availability. Access Complexity indicates how difficult it is to exploit the vulnerability once the attacker has gained access. It is defined in terms of three levels: High, Medium, and Low. Authentication indicates whether

[30]https://cloud.google.com/translate/docs

authentication is required by the attacker to exploit the vulnerability. It is a binary identifier taking the values Required and Not Required. Confidentiality, Integrity, and Availability indicate what loss the system would incur if the vulnerability is exploited. They take the values None, Partial, and Complete. All the CVSS vector features are categorical. These features are vectorized by building a vector with all possible categories. Then if that category is present, 1 is inserted, otherwise 0.

4.4.2.3 Language

DW feeds are posted in different languages. The four most frequently used languages in DW posts that describe vulnerabilities are English, Chinese, Russian, and Swedish. Since there is limited number of non-English postings, giving the model little chance to learn adequate representation for each language, text translation was used, as mentioned earlier. To this end, while translation can result in a loss of important information, the impact can be minimized by using the language as a feature. The analysis on the languages of DW feeds and their variation in the exploitation rate are detailed in Sect. 4.5.

4.4.2.4 Presence of Proof-of-Concept

The presence of PoC exploits in EDB increases the likelihood of vulnerability exploitation. It can be treated as a binary feature that indicates whether a PoC is present for a vulnerability or not.

4.4.2.5 Vulnerability Mention on ZDI

ZDI acts similar to the NVD, i.e., both disclose software vulnerabilities. Given that a vulnerability is disclosed on ZDI, its exploitation likelihood raises. Similar to the presence of PoCs, a binary feature can be used to denote whether a vulnerability was disclosed in ZDI before it is exploited.

4.4.2.6 Vulnerability Mention on the Dark Web or Deep Web

A binary feature can be used to indicate whether a vulnerability is mentioned on the Dark Web or the Deep Web.

4.5 Vulnerability and Exploit Analysis

To assess the importance of aggregating different data sources for early identification of threatened vulnerabilities, first the likelihood of exploitation is analyzed provided that a vulnerability is mentioned by each data source. Time-based analysis is then performed for the exploited vulnerabilities that have reported exploitation dates ($n = 194$) to show the difference in days between when vulnerabilities are exploited and when they are mentioned online. Our time-based analysis ignores the exploited vulnerabilities without reported dates, because no assumption can be made about their exploitation dates.

To identify the vendors of highly vulnerable software and systems, the ground truth has to be analyzed and compared to other sources. As previous works described, Symantec reports attack signatures for vulnerabilities of certain products [4, 8]. To show the variation in vendor coverage attained from various data sources, the distribution of the target software vendors are studied per data source. This analysis is based on the vendor mentions by CPE data from the NVD. Note that a vulnerability can appear in various software versions, including versions developed for different platforms, and therefore a single vulnerability can potentially be mapped to multiple software vendors. Finally, a language-based analysis is provided on Deep Web/Dark Web data to reveal some socio-cultural factors present in the Deep Web and the Dark Web that seem to affect the likelihood of exploitation.

4.5.1 Likelihood of Exploitation

For each data source, Table 4.3 shows the vulnerability exploitation probability for the vulnerabilities mentioned in that data source. This analysis emphasizes the value of open data sources in supplementing the NVD data. As mentioned in Sect. 4.4.1, approximately 2.4% of the vulnerabilities in NVD are exploited in the wild. Hence, including other sources can increase the likelihood of correctly identifying the vulnerabilities that will be exploited.

Table 4.3 Number of vulnerabilities, number of exploited vulnerabilities, fraction of exploited vulnerabilities that appeared in each source, and fraction of total vulnerabilities that appeared in each source. Results are reported for vulnerabilities and exploited vulnerabilities appeared in EDB, ZDI, DW (distinct CVEs), CVEs in ZDI or DW, and CVEs in any of the three sources

	EDB	ZDI	DW	ZDI/DW	EDB/ZDI/DW
Number of vulnerabilities	799	824	378	1,180	1,791
Number of exploited vulnerabilities	74	95	52	140	164
Percentage of exploited vulnerabilities	21%	31%	17%	46%	54%
Percentage of total vulnerabilities	6.3%	6.5%	3.0%	9.3%	14.2%

4.5.2 *Time-Based Analysis*

Most software systems are attacked repeatedly using vulnerabilities after exploits to such vulnerabilities have been detected in the wild [29]. As a matter of fact, a long time may pass between the date the vulnerabilities are disclosed and the date they are patched by vulnerability management teams. Here, only the population of exploited vulnerabilities that are reported with their exploitation date are analyzed (194 vulnerabilities).

Figure 4.3 shows that for more than 93% of the vulnerabilities, they are disclosed by NIST before any real-world attacks are detected. In the other few cases, attacks were detected in the wild before NIST published the vulnerabilities (i.e., zero-day attacks). This could be caused by (1) the vulnerability information is sometimes leaked before the disclosure, (2) by the time NIST disclosed a vulnerability in the NVD, other sources have already validated and published it, then exploits rapidly started using it in real-world attacks, or (3) the attacker knew that what they were doing was successful and continued to exploit their targets until discovered [2]. Additionally, ZDI and NVD have limited variation on the vulnerability disclosure dates (median is 0 days). It is important to note that because ZDI disclosures come from the industry, reserved CVE numbers are shown earlier here than in other sources.

In the case of EDB, almost all of the exploited vulnerabilities that have PoCs archived were found in the wild within the first 100 days of the PoCs availability. Such a short time period between the availability of PoCs and actual real-world attacks indicates that having a template for exploits (in this case PoCs) makes it easy for hackers to configure and use in real-world attacks. Figure 4.4 shows the difference in days between the availability of PoCs and exploitation dates.

In the case of the DW database, more than 60% of the first-time mentions to the exploited vulnerabilities are within 100 days before or after the exploitation dates.

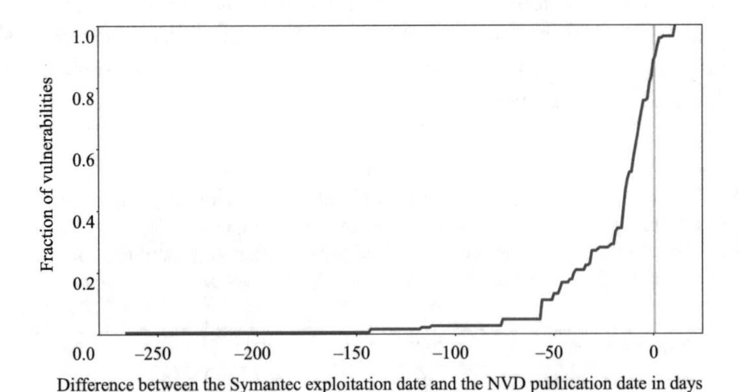

Fig. 4.3 Day difference between the CVE first published in the NVD and the Symantec attack signature date versus the fraction of exploited CVEs the NVD reported (cumulative)

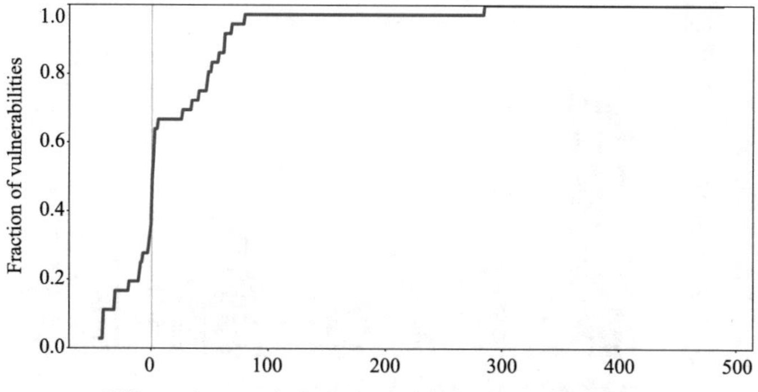

Fig. 4.4 Day difference between the date of availability of PoC on EDB and the Symantec attack signature date versus the fraction of exploited CVEs with PoCs EDB reported (cumulative)

The remaining mentions are within the 18 months time frame after the vulnerability exploitation date (see Fig. 4.5).

4.5.3 *Vendor-/Platform-Based Analysis*

As noted, Symantec reports vulnerabilities that attack the systems and software configurations used by their customers. For the studied vulnerabilities, more than 84% and 36% of the exploited vulnerabilities reported by Symantec exist in products solely from, or run on, Microsoft and Adobe products, respectively; whereas less than 16% and 8% of vulnerabilities published in the NVD are related to Microsoft and Adobe, respectively. Figure 4.6 shows the percentage from the exploited vulnerabilities that

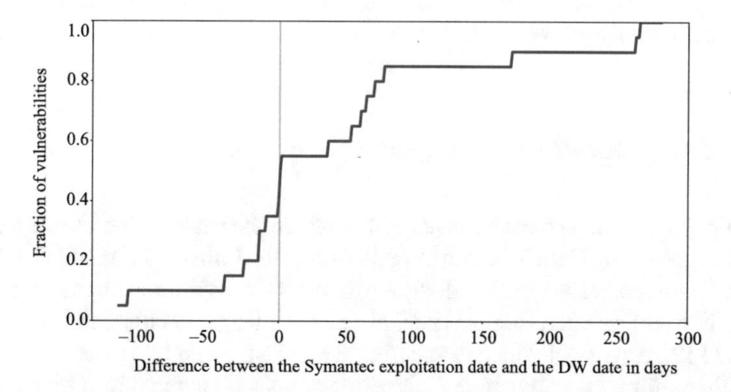

Fig. 4.5 Day difference between CVE first mentioned in DW and Symantec attack signature date versus the fraction of exploited CVEs reported on DW (cumulative)

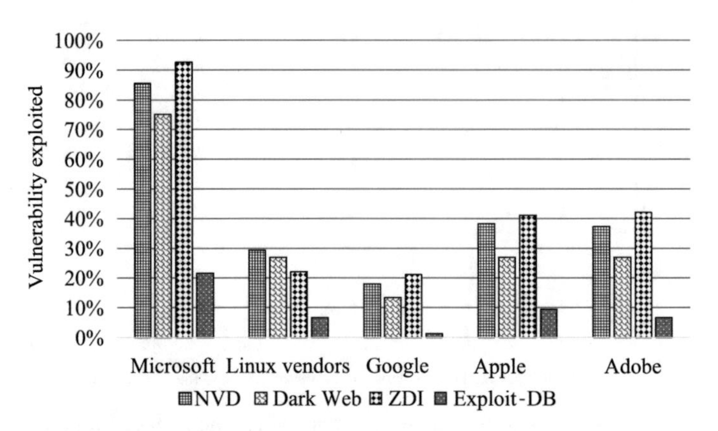

Fig. 4.6 Most exploited vendors in each data source

can affect each of the top five vendors in every data source. It is important to note that a vulnerability may affect more than a single vendor (e.g., CVE-2016-4272 exists in Adobe Flash Player,[31] and it allows attackers to execute arbitrary code via unspecified vectors and can affect products from all five vendors). Additionally, the absence of vulnerabilities detected in other important systems and software vendors from Symantec's dataset does not imply that they have not been exploited; rather, they are not detected by Symantec (i.e., false negatives). Furthermore, the presence of some operating systems (e.g., Linux) in the exploited vulnerabilities does not necessarily imply good coverage of Symantec data to these systems; however, other exploited products can run on these operating systems.

Furthermore, DW data appears to have more uniform vendor coverage. Only 30% and 6.2% of the vulnerabilities mentioned in DW during the time period of this study were related to Microsoft and Adobe, respectively. Additionally, ZDI favors products from these two vendors (57.8% in the case of Microsoft and 35.2% in the case of Adobe). This provides evidence that each data source cover vulnerabilities that target various sets of software vendors.

4.5.4 Language-Based Analysis

Interestingly, notable variations have been found on the exploitation likelihood w.r.t. the language used on Dark Web and Deep Web data feeds that reference CVEs. In DW feeds, four languages are detected with different vulnerability post and item distributions. Not surprisingly, English and Chinese have far more vulnerability mentions (242 and 112, respectively) than Russian and Swedish (13 and 11, respectively). However, vulnerabilities mentioned in Chinese postings are characterized by the lowest

[31] https://www.adobe.com/products/flashplayer.html

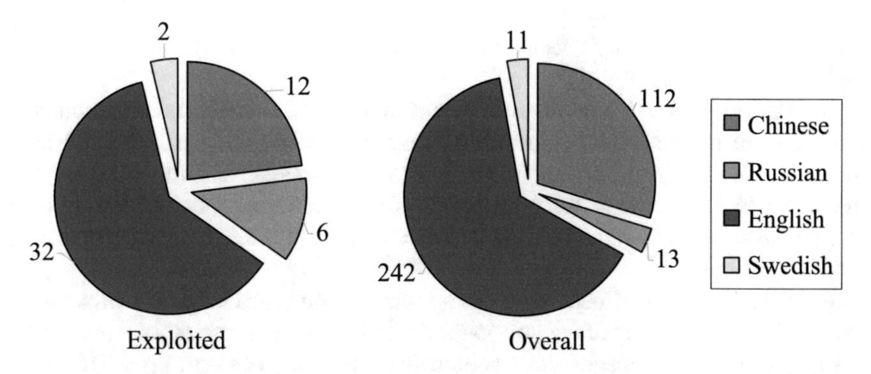

Fig. 4.7 The number of *exploited* vulnerabilities mentioned by each language (left), and the number of vulnerabilities mentions in each language (right)

exploitation rate. For example, of those vulnerabilities, only 12 are exploited (about 10%), while 32 of the vulnerability mentioned on English postings are exploited (about 13%). Although vulnerability mentions in Russian or Swedish postings are few, these vulnerabilities have a very high exploitation rate. For example, about 46% of the vulnerabilities mentioned in Russian were exploited (6), and about 19% for vulnerabilities mentioned in Swedish (2). Figure 4.7 shows the number of vulnerability mentions by each language as well as the number of exploited vulnerabilities mentioned by each language.

4.6 Experimental Setup

To evaluate our approach, a series of experiments was conducted with the presented models. First, our model was compared to a benchmark work presented in [8]. For our model, random forest gave the best F1 measure.[32] Random forest is an ensemble method proposed by Breiman [30], which is based on the idea of generating multiple predictors (decision trees in this case) to be used in combination to classify a new disclosed vulnerability. The strength of the random forest lies in introducing randomness to build each classifier and using random, low-dimensional subspaces to classify the data at each node in a classifier. A random forest that combines bagging [30] for each tree with random feature selection [31] at each node to split the data was used. The final result was therefore an ensemble of decision trees, each having their own independent opinion on class labels (i.e., exploited or not exploited for a given disclosed vulnerability). Therefore, new vulnerabilities were classified independently by each tree, and the most suitable class label was assigned to them. Multiple decision trees may result in having multiple class labels for the same data

[32]The harmonic mean of precision and recall.

sample; hence, a majority vote was taken and the class label with most votes was assigned to the vulnerability.

In [8], the authors temporally mix their samples, i.e., vulnerabilities exploited in the future are used to predict vulnerabilities exploited in the past, a practice discussed in [14]. Additionally, to account for the severe class imbalance, only vulnerabilities that occur in Microsoft or Adobe products were used in training and testing (477 vulnerabilities, 41 of which were exploited). Our model is compared to [8] under the same conditions.

For the second experiment, the training samples are restricted to the vulnerabilities published before any of the vulnerabilities in the testing samples were published. Also, only data feeds that are present before the exploitation date are used. This guarantees that the experimental settings resemble the real world. Because no assumptions can be made regarding the sequence of events for the exploited vulnerabilities reported by Symantec without the exploitation date ($n = 112$), these vulnerabilities are removed from the experiments. The fraction of exploited vulnerabilities becomes 1.2%. The performance of our model is compared to the CVSS score.

Additionally, the goal for exploit prediction is to predict whether a disclosed vulnerability will likely be exploited in the future or not. Few vulnerabilities are exploited before they are published [2]. Prediction for such vulnerabilities does not add any value to the goal of the prediction task, considering that they had been already exploited by the time those vulnerabilities are revealed. That being said, knowing what vulnerabilities are exploited in the wild can help organizations with their cyber defense strategies, but this is out of the scope of this chapter.

4.6.1 Performance Evaluation

Our classifiers are evaluated based on two classes of metrics that have been used in previous work. The first class is used to demonstrate the performance achieved on the minority class (in our case 1.2%). The metrics under this class are precision and recall. They are computed as shown in Table 4.4. Precision is defined as the fraction of vulnerabilities that were exploited from all vulnerabilities predicted to be exploited by our model. It highlights the effect of mistakingly flagging non-exploited vulnerabilities. Recall is defined as the fraction of correctly predicted exploited vulnerabilities from the total number of exploited vulnerabilities. It highlights the effect of unflagging important vulnerabilities that were used later in attacks. For highly imbalanced data, these metrics give us an intuition regarding how well the classifier performed on the minority class (i.e., exploited vulnerabilities). The F1 measure summarizes precision and recall in a common metric. It may vary depending on the trade-off between precision and recall, which it turn depends on the application. If keeping the number of incorrectly flagged vulnerabilities to a minimum is a priority, then high precision is desired. To keep the number of undetected vulnerabilities that are later exploited to a minimum, high recall is desired. The *Receiver Operating Characteristics* (ROC) curve and the *Area Under Curve* (AUC) of the classifier

Table 4.4 Evaluation metrics.
TP—true positive, FP—false positive, FN—false negative, TN—true negative

Metric	Formula
Precision	$\frac{TP}{TP+FP}$
TPR (recall in case of binary classification)	$\frac{TP}{TP+FN}$
F1	$2 \cdot \frac{precision \cdot recall}{precision+recall}$
FPR	$\frac{FP}{FP+TN}$

are also described. ROC visualizes the performance of our classifier by plotting the true positive rate against the false positive rate at various thresholds of the confidence scores the classifier outputs. In binary classification problems, the overall TPR is always equivalent to recall for the positive class, while FPR is the number of non-exploited vulnerabilities that are incorrectly classified as exploited from all the samples that have not been exploited. ROC is a curve, thus, AUC is the area under ROC. The higher the AUC value, the more optimal the model (i.e., a classifier with an AUC of 1 is a perfect classifier).

4.6.2 *Results*

The performance of several standard supervised machine learning approaches were used and compared for exploit prediction. Parameters for all approaches were set in a way to provide the best performance. The *scikit-learn*[33] Python package [32] was used.

4.6.2.1 Examining Classifiers

The temporal information is maintained for all the classifiers. The vulnerabilities are sorted by the date they were posted on NVD. The first 70% are reserved for training, along with any features that are available by the end of the training period. The remaining vulnerabilities are used for testing.

Table 4.5 shows a comparison between the classifiers with respect to precision, recall and F1 measure.

Random forest (RF) performs the best with an F1 measure of 0.4 as compared to Support Vector Machine's 0.34, Bayesian Network's 0.34, Logistic Regression's 0.33, Decision Tree's 0.25, and Naïve Bayes's 0.27. Note that even though RF has the best F1 measure, it does not have the best recall—NB does. RF with high precision is chosen, which makes the model reliable compared to low precision, which results in a lot of false positives.

[33]http://scikit-learn.org

Table 4.5 Precision, recall, and F1 measure for RF, SVM, LOG-REG, DT and NB to predict whether a vulnerability will likely be exploited

Classifier	Precision	Recall	F1 measure
RF	**0.45**	0.35	**0.40**
BN	0.31	0.38	0.34
SVM	0.28	0.42	0.34
LOG-REG	0.28	0.4	0.33
DT	0.25	0.24	0.25
NB	0.17	**0.76**	0.27

4.6.2.2 Benchmark Test

Our model is compared to a recent work that uses vulnerability mentions on Twitter to predict the likelihood of exploitation [8]. In this study, the authors used SVM as the classifier, while our model works best with a random forest classifier. Although our approach is expected to achieve better performance, a comparison with this work is important due to two reasons: (1) to the best of our knowledge, there is no existing work on predicting exploits in the wild using Deep Web/Dark Web data, and (2) the comparison between all major approaches confirmed that using feeds from social media is currently the best approach.

In [8], the authors restricted the training and evaluation of their classifier to vulnerabilities targeting Microsoft and Adobe products, because Symantec does not have attack signatures for all the targeted platforms. They performed a 10-fold stratified cross-validation, where the data is partitioned into 10 parts while maintaining the class ratio in each part. They trained on 9 parts and tested on the remaining one. The experiment was repeated for all 10 parts. Hence, each sample was tested once.

For comparison, the same experiment is performed under highly similar assumptions. All exploited vulnerabilities are used regardless of whether the date is reported by Symantec. In our case, there were 2,056 vulnerabilities targeting Microsoft and Adobe products. Out of these, 261 were exploited, which is consistent with previous works (by rate). A 10-fold stratified cross-validation is performed. The precision-recall curve of this model is depicted in Fig. 4.8.

The precision-recall curve shows the trade-off between precision and recall for different decision threshold. Since the F1 measure is not reported in [8], the reported precision-recall curve is used for comparison while maintaining the recall value constant. Table 4.6 shows the comparison between the two models in terms of precision for different values of recall.

For a threshold of 0.5, the F1 measure is 0.44 with precision 0.53 and recall 0.3. Maintaining the recall, the precision displayed in the graph in [8] is 0.3, which is noticeably lower than 0.4. To compare the precision, the same experiment is performed on different recall values. At each point, a higher precision is obtained than with the previous approach.

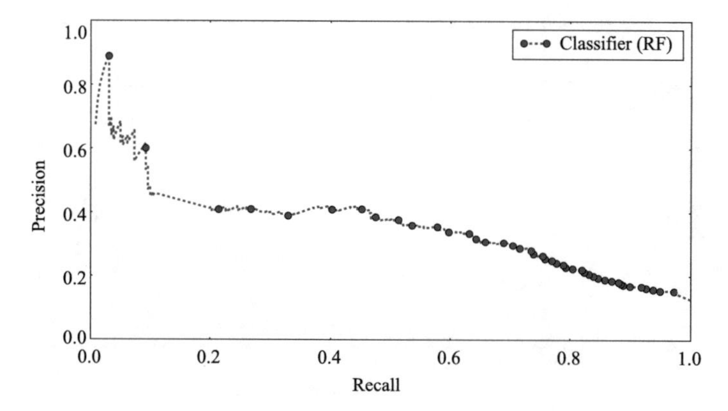

Fig. 4.8 Precision-recall curve for proposed features for Microsoft-Adobe vulnerabilities (RF)

Table 4.6 Precision comparison between [8] and the proposed model while keeping the recall constant

Recall	Precision[a]	Precision (our work)
0.20	0.30	**0.41**
0.40	0.18	**0.40**
0.70	0.10	**0.29**

[a]Numbers derived from Fig. 6a from [8]

4.6.2.3 Baseline Comparison

Bullough et al. argue that the problem of predicting the likelihood of exploitation is sensitive to the sequence of vulnerability-related events [14]. Temporally mixing such events leads to future events being used to predict past ones, resulting in inaccurate prediction results. To avoid the temporal mixing of events, time-based splits are created, as described in this section.

For baseline comparison, the CVSS version 2.0 base score is used to classify whether a vulnerability will likely be exploited based on the associated severity score. CVSS score has been used as a baseline in previous studies [4, 8]. CVSS tends to be overly cautious, i.e., it tends to assign high scores to many vulnerabilities, resulting in many false positives. Figure 4.9 shows the precision-recall curve for the CVSS score.

It is computed by varying the decision threshold (x-axis), based on which the class label of each vulnerability can be determined. CVSS gives high recall with very low precision, which is not desired for real-world patch prioritization tasks. The best F1 measure that could be obtained is 0.15. Figure 4.10 shows the performance comparison between our proposed RF model and the CVSS-based model that yields the highest F1 score.

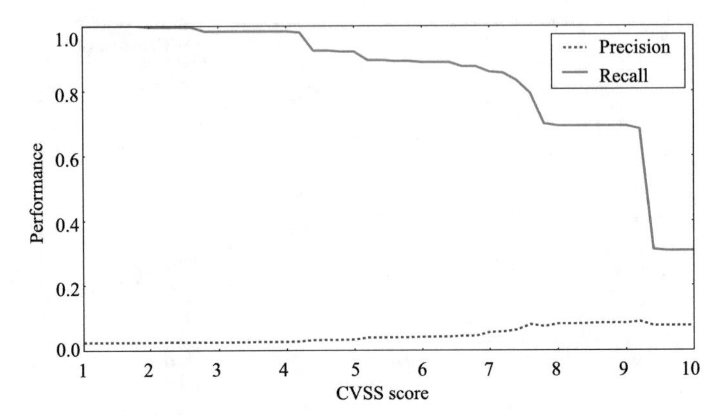

Fig. 4.9 Precision and recall for classification based on CVSS base score version 2.0 threshold

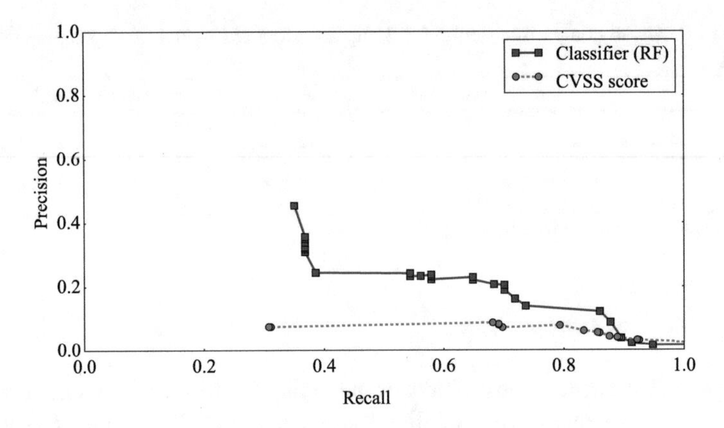

Fig. 4.10 Precision-recall curve for classification based on CVSS score threshold (RF)

Our model outperforms the baseline with an F1 measure of 0.4, a precision of 0.45, and a recall of 0.35. Additionally, our classifier shows very high TPR (90%) at low FPR (13%), with an AUC of 94%, as depicted in Fig. 4.11.

4.6.2.4 Evaluation with Individual Data Sources

Our study investigated how the prediction of vulnerabilities mentioned in a data source is affected by introducing the corresponding data source. This is important for understanding if the addition of a particular data source benefits the vulnerabilities that have been mentioned in that data source. It turned out that time-based split used in the previous experiments leaves very few vulnerabilities mentioned in these data sources in the test set (ZDI: 18, DW: 4, EDB: 2). Therefore, the numbers were incresed by (1) performing a 10-fold cross-validation without sorting the vulnerabilities, and (2) increasing the ground truth by considering the exploited vulnerabilities that did

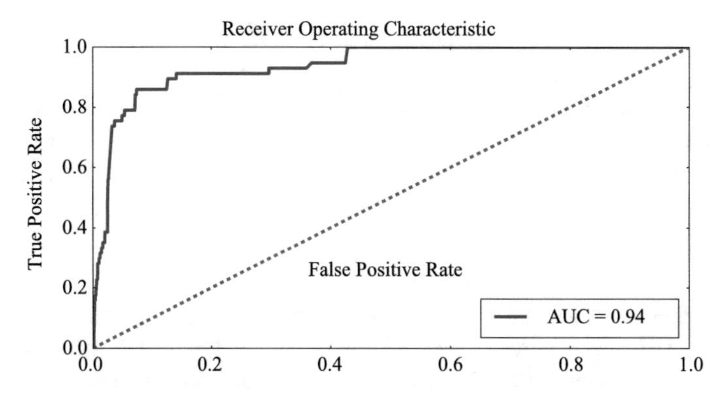

Fig. 4.11 ROC curve for classification based on a random forest classifier

not have exploit date (these were removed from earlier experiments because it was not clear whether these were exploited before or after the vulnerability was exploited). Using these two techniques, there were 84 vulnerabilities mentioned in ZDI that have been exploited, 57 in EDB, and 32 in DW. The results (precision, recall, and F1) for the vulnerabilities mentioned in each data source were captured. Also, the prediction of these vulnerabilities are described by using only the NVD features. For the vulnerabilities mentioned in DW, only DW features provided along with the NVD features are considered. The model predicts 12 vulnerabilities as exploited with a precision of 0.67 and recall of 0.38. By only considering the NVD features, the model predicts 12 vulnerabilities as exploited with a precision of 0.23 and recall of 0.38. Consequently, using the DW features, the precision improved significantly from 0.23 to 0.67. Table 4.7 shows the precision-recall with the corresponding F1 measure. DW information was thus able to correctly identify positive samples mentioned in DW with higher precision.

For ZDI, 84 vulnerabilities were mentioned. Using only NVD features yields to an F1 measure of 0.25 (precision: 0.16, recall: 0.54) as compared to adding the ZDI feature with an F1 measure of 0.32 (precision: 0.49, recall: 0.24)—a significant improvement in precision. Table 4.7 also shows the precision-recall with the corresponding F1 measure for samples mentioned on ZDI.

Table 4.7 Precision, recall, and F1 measure for vulnerabilities mentioned on DW, ZDI, and EDB

Source	Case	Precision	Recall	F1 measure
DW	NVD	0.23	0.38	0.27
	NVD + DW	0.67	0.375	0.48
ZDI	NVD	0.16	0.54	0.25
	NVD + ZDI	0.49	0.24	0.32
EDB	NVD	0.15	0.56	0.24
	NVD + EDB	0.31	0.40	0.35

Table 4.8 Performance improvement attained by applying SMOTE for the BN classifier using different oversampling for the exploited samples

Oversampling [%]	Precision	Recall	F1 measure
100	0.37	0.42	0.39
200	0.40	**0.44**	**0.42**
300	**0.41**	0.40	0.40
400	0.31	0.40	0.35

Similar analysis was performed for the vulnerabilities that had PoCs available on EDB. For EDB, there were 57 vulnerabilities with PoCs. When using only NVD features, this time the model scored an F1 measure of 0.24 (precision: 0.15, recall: 0.56); while adding the EDB feature boosted the F1 measure to 0.35 (precision: 0.31, recall: 0.40), which is a significant improvement in precision, as shown in Table 4.7.

4.6.2.5 Feature Importance

To better explain our feature choices, an explanation of where our prediction power primarily derives from is due. The features that have the most contribution to the prediction performance are reported. A feature vector for a sample has 28 features computed from the non-textual data (summarized in Table 4.2) as well as the textual features—TF-IDF computed from the bag of words for the 1,000 words that have the highest frequency in the NVD description and DW. For each feature, the Mutual Information (MI) [33] is computed, which expresses how much a variable (here a feature x_i) tells about another variable (here the class label $y \in \{exploited, not\ exploited\}$). The features that contribute the most from the non-textual data are $\{language\_Russian = true, has\_DW = true, has\_PoC = false\}$. In addition, the features that contribute the most from the textual data are the words {*demonstrate, launch, network, xen, zdi, binary, attempt*}. All of these features received MI scores higher than 0.02.

4.6.2.6 Addressing Class Imbalance

The problem of class imbalance has gained lot of research interest [34]. Since our dataset is highly imbalanced, SMOTE [28] is used and the improvement is measured in terms of classification performance. SMOTE oversamples the minority class by creating synthetic samples with features similar to those of the exploited vulnerabilities. This data manipulation is only applied to the training set. Using SMOTE, no performance improvement is achieved for our RF classifier. However, SMOTE introduces a considerable improvement with a Bayesian Network classifier. Table 4.8 reports different oversampling ratios and the change in performance. The best oversampling ratio is experimentally determined, i.e., high oversampling ratios lead to the model learning from a distribution that differs significantly from the real distribution.

4.7 Adversarial Data Manipulation

The effects of adversarial data manipulation is studied here only on DW data. As for the presence of PoCs, only those PoCs that are verified by EDB are considered. Adversaries need to hack into EDB to add noise or remove PoCs from the EDB database. In this analysis, it is assumed that such an action cannot be taken by adversaries. Similarly, ZDI publishes only vulnerabilities that are verified by its researchers; hence there is a very small chance of manipulating these data sources.

On the other hand, the public nature of DW marketplaces and forums gives an adversary the ability to poison the data used by the classifier. They can achieve this by adding vulnerability discussions on these platforms with the intent of fooling the classifier to make it produce high false positives. Previous works discuss how an adversary can influence a classifier by manipulating the training data [35–37].

In our prediction model, the presence of the vulnerability in DW is used, along with the language of the market/forum on which it was mentioned, and the vulnerability description, as features. An adversary could easily post discussions regarding vulnerabilities he does not intent to exploit, nor does he expect that they will be exploited. To study the influence of such noise on the performance of the model, experiments were conducted using the following two strategies:

1. **Adversary adding random vulnerability discussions**. In this strategy, the adversary initiates random vulnerability discussions on DW and reposts them with a different CVE number. So the CVE mentions on DW increases. For our experiment, two cases are considered with different amounts of noise added. In case 1, it is assumed that the noise is present in both the training and the testing data. Various fractions of noise are considered (5, 10, 20% of the total data samples), randomly distributed in training and testing data. The experimental setup follows conditions discussed in Sect. 4.6. Vulnerabilities are first sorted according to time, and the first 70% are reserved for training and the remaining for testing. Figure 4.12a shows the ROC curve showing the false positive rate (FPR) versus the true positive rate (TPR). For different amount of noise introduced, our model still maintains a high TPR with low FPR and an AUC of at least 0.94, a performance similar to the experiment without adding noise. This shows that the model is highly robust against noise such that it learns good representation of the noise in the training dataset and can distinguish them in the testing dataset.
   For case 2, random vulnerability discussions found on DW are added with a different CVE number only to the test data and repeat the same experiment. Figure 4.12a shows the ROC plot. In the case, even though there is a slight increase in the FPR, the performance is still on par with the experiment without noise (AUC $\geqslant 0.87$). Hence, noisy sample affect the prediction model slightly, if no noise was introduced in the training data.
2. **An adversary adds a vulnerability discussion similar to the NVD description**. In the previous strategy, the adversary adds vulnerability discussions randomly without taking into account the actual capability of the vulnerability. For instance,

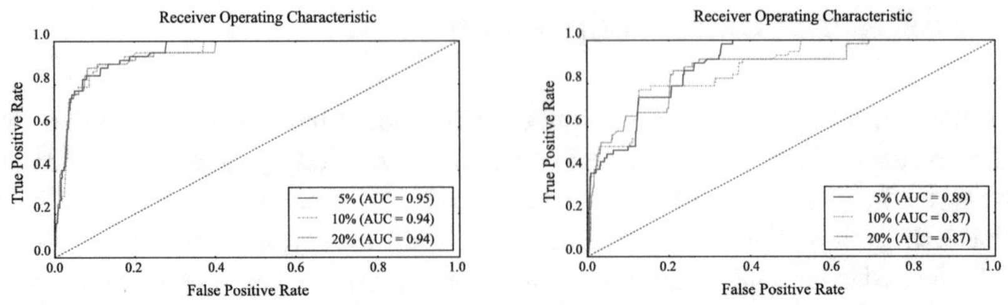

(a) ROC curves for adversary adding noise to both the training and the testing data (left), and only adding noise to the testing data (right) (Strategy 1)

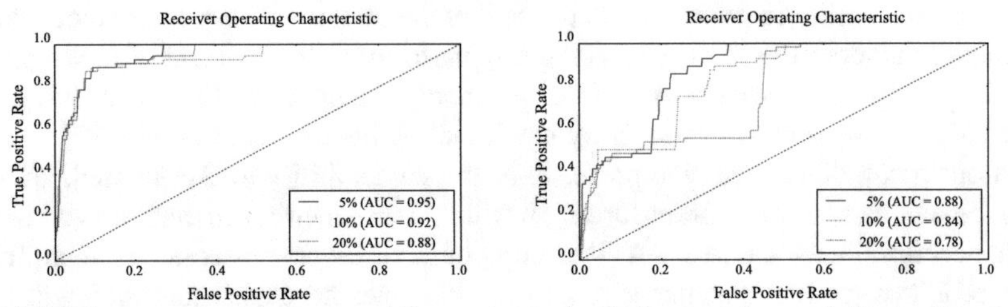

(b) ROC curves for adversary adding noise to both the training and the testing data (left), and only adding noise to the testing data (right) (Strategy 2)

Fig. 4.12 ROC curves demonstrating the robustness of the random forest model against adversarial data manipulation

CVE-2016-3350[34] is a vulnerability in Microsoft Edge, as reported by the NVD. If the vulnerability is mentioned on DW as noise by an adversary but targeting Google Chrome, then it might be easy for the prediction model to detect it as seen in previous experiments. But, what if the adversary crafts the vulnerability discussion such that it is a copy of the NVD description or consistent with the NVD description? In this strategy, the adversary posts the NVD description with the CVE number on DW. For case 1, this noise is considered to be randomly distributed in both training and testing. Figure 4.12b shows the ROC curves for different levels of noise. The performance decreases as the number of noisy samples increases, but there is no significant decline (AUC $\geqslant$ 0.88).

The experiment was repeated by adding noise only to the test data for case 2. This experiment indicated that the biggest drop in performance results in an AUC of 0.78, when 20% of the samples are noise (see Fig. 4.12b). This confirms that adding correct vulnerability discussions does affect the prediction model, except if a large number of such samples are added. Also, the effect can be countered by also adding such noisy samples to the training data for the model to learn from.

Note that an adversary would need to add a large number of noisy samples to lower the performance of the prediction model. Previous research on using data feeds like Twitter for exploit prediction mentions that an adversary can register a

[34]https://nvd.nist.gov/vuln/detail/CVE-2015-3350

large number of Twitter accounts and flood Twitter with vulnerability mentions [8]. In DW markets and forums, creation of accounts needs verification and, in some cases, technical skills. While fake accounts are often sold on the dark web, it is difficult to purchase and maintain thousands of such fake accounts to post with them. Also, if someone posts a large volume of discussions with CVE mentions, he can be identified from his username or can be removed from the market/forum if many of his posts get downvoted for being irrelevant. It is also important to note that such forums also function as a meritocracy [19], where users who contribute more are held in higher regard (which also makes it difficult to flood discussions with such information).

4.8 Discussion

Viability of the model and cost of misclassification. The performance achieved by our model as a first-line defense layer is very promising. Recall that a random forest classifier outputs a confidence score for every testing sample. A threshold can be set to identify the decision boundary. Note that all the reported results were based on hard-cut thresholds, such that all samples that are assigned confidence scores greater than a threshold thr are predicted to be exploited. Relying solely on a hard-cut threshold may not be a good practice in real-world threat assessments; rather, thr should be varied in accordance to other variables within an organization such that different thresholds can be set to different vendors (i.e., thr_{ven1}, thr_{ven2}), or information systems (i.e., thr_{sys1}, thr_{sys2}). For instance, if an organization hosts an important website on an Apache server, and the availability of this site is of top priority for the organization, then any vulnerability of the Apache server should receive high attention and put forward to remediation plans regardless of other measures. Other vulnerabilities, tens of which are disclosed on a daily bases, may exist in many other systems within the organization.

Since it is very expensive to be responsive to that many security advisories (e.g., some patches may be unavailable, some systems may need to be taken offline to apply patches), assessing the likelihood of exploitation can help in quantifying the risk and planning mitigations. Risk is always thought of as a function of likelihood (exploitation) and impact. The cost of classifying negative samples as *exploited* is the effort made to have them fixed. This mostly involves patching or other remediations such as controlling access or blocking network ports. Similarly, the cost of misclassification depends on the impact. For example, if two companies run the same database management system, and one hosts a database with data about all business transactions, while the other hosts a database with data that is of little value to the company, the resulting cost of a data breach would be significantly different.

Model success and failure cases. The analysis of false negatives and false positives provides an understanding of why and in which case our model performs well and when is it inefficient. The 10 exploited vulnerabilities (about 18% of the exploited samples in the testing dataset) that received the lowest confidence scores seem to have common features. For example, 9 of these appear in Adobe products, namely Flash Player (five vulnerabilities) and Acrobat Reader (four vulnerabilities). Flash Player vulnerabilities seem to have very similar description in the NVD. The same holds for Acrobat Reader. They were assigned CVE-IDs on the same day (April 27, 2016), and 7 out of these 9 were published on the same day as well (July 12, 2016), and assigned a CVSS base score of 10.0 (except for one, assigned 7.0). The other vulnerability exist in Windows Azure Active Directory (CVSS score of 4.3). Out of these 10 vulnerabilities, one had a verified PoC archived on EDB before it was detected in the wild, and another one had a ZDI mention, while none were mentioned in DW. The reason for misclassifying these vulnerabilities is the limited representation of these samples in the training dataset. Moreover, this observation signifies the importance of avoiding experiments on time-intermixed data, a point discussed in Sect. 4.3.2.

False positive samples were also studied, which received high confidence score—samples our model predicted as exploited while they were not. For our random forest classifier, all the examined false positives appeared in Microsoft products, although vendor was not used as a feature. Our model is able to infer the vendor from other textual features. Assumingly, this level of overfitting is inevitable and marginal, and is caused mainly by the limitations of the ground truth. The model is highly generalizable though. There are examples of vulnerabilities from other vendors with confidence scores close to the thr used by us; however, it cannot be assumed that these vulnerabilities are exploited.

4.9 Conclusion

This chapter proposed an approach that aggregates early signs of vulnerability exploitation from various online sources to predict the likelihood of exploitation, a problem directly related to patch prioritization. By conducting a series of experiments, the use of cyberthreat feeds from the data sources was demonstrated. Our machine learning model outperforms existing models that combine information from social media sites like Twitter for exploit prediction. More precisely, our results show that while maintaining a high true positive rate, a reasonably low false positive rate can be achieved in predicting exploits compared to previous severity scoring systems.

Acknowledgements Some of the authors were supported by the Office of Naval Research (ONR) contract N00014-15-1-2742, the Office of Naval Research (ONR) Neptune program and the ASU Global Security Initiative (GSI). Paulo Shakarian and Jana Shakarian are supported by the Office of the Director of National Intelligence (ODNI) and the Intelligence Advanced Research Projects Activity (IARPA) via the Air Force Research Laboratory (AFRL) contract number FA8750-16-C-0112. The U.S. Government is authorized to reproduce and distribute reprints for Governmental purposes notwithstanding any copyright annotation thereon. Disclaimer: The views and conclusions contained herein are those of the authors and should not be interpreted as necessarily representing the official policies or endorsements, either expressed or implied, of ODNI, IARPA, AFRL, or the U.S. Government.

References

1. Pfleeger CP, Pfleeger SL, Margulies J (2015) Security in computing, 5th edn. Prentice Hall, Upper Saddle River, NJ, USA
2. Bilge L, Dumitras T (2012) Before we knew it: an empirical study of zero-day attacks in the real world. In: Yu T, Danezis G, Gligor V (eds) Proceedings of the 2012 ACM Conference on Computer and Communications Security. ACM, New York, pp 833–844. https://doi.org/10.1145/2382196.2382284
3. Frei S, Schatzmann D, Plattner B, Trammell B (2010) Modeling the security ecosystem–The dynamics of (in)security. In: Moore T, Pym D, Ioannidis C (eds) Economics of information security and privacy. Springer, Boston, pp 79–106. https://doi.org/10.1007/978-1-4419-6967-5_6
4. Allodi L, Massacci F (2014) Comparing vulnerability severity and exploits using case-control studies. ACM Trans Inform Syst Secur 17(1), Article No. 1. https://doi.org/10.1145/2630069
5. Durumeric Z, Kasten J, Adrian D, Halderman JA, Bailey M, Li F, Weaver N, Amann J, Beekman J, Payer M, Weaver N, Adrian D, Paxson V, Bailey M, Halderman JA (2014) The matter of Heartbleed. In: Williamson C, Akella A, Taft N (eds) Proceedings of the 2014 Conference on Internet Measurement Conference. ACM, New York, pp 475–488. https://doi.org/10.1145/2663716.2663755
6. Edkrantz M, Said A (2015) Predicting cyber vulnerability exploits with machine learning. In: Thirteenth Scandinavian Conference on Artificial Intelligence, pp 48–57. https://doi.org/10.3233/978-1-61499-589-0-48
7. Nayak K, Marino D, Efstathopoulos P, Dumitraş T (2014) Some vulnerabilities are different than others. In: Stavrou A, Bos H, Portokalidis G (eds) Research in attacks, intrusions and defenses. Springer, Cham, pp 426–446. https://doi.org/10.1007/978-3-319-11379-1_21
8. Sabottke C, Suciu O, Dumitras T (2015) Vulnerability disclosure in the age of social media: exploiting Twitter for predicting real-world exploits. In: Proceedings of the 24th USENIX Security Symposium. USENIX Association, Berkeley, CA, USA, pp 1041–1056. https://www.usenix.org/sites/default/files/sec15_full_proceedings.pdf
9. Allodi L, Massacci F (2012) A preliminary analysis of vulnerability scores for attacks in wild: the EKITS and SYM datasets. In: Yu T, Christodorescu M (eds) Proceedings of the 2012 ACM Workshop on Building Analysis Datasets and Gathering Experience Returns for Security. ACM, New York, pp 17–24. https://doi.org/10.1145/2382416.2382427
10. Mittal S, Das PK, Mulwad V, Joshi A, Finin T (2016) CyberTwitter: using Twitter to generate alerts for cybersecurity threats and vulnerabilities. In: Subrahmanian VS, Rokne J, Kimar R, Caverlee J, Tong H (eds) Proceedings of the 2016 IEEE/ACM International Conference on Advances in Social Networks Analysis and Mining. IEEE Press, Piscataway, NJ, USA, pp 860–867
11. Marin E, Diab A, Shakarian P (2016) Product offerings in malicious hacker markets. In: Zhou L, Kaati L, Mao W, Wang GA (eds) Proceedings of the 2016 IEEE Conference on Intelligence and Security Informatics. The Printing House, Stoughton, WI, USA, pp 187–189. https://doi.org/10.1109/ISI.2016.7745465
12. Samtani S, Chinn K, Larson C, Chen H (2016) AZSecure hacker assets portal: cyber threat intelligence and malware analysis. In: Zhou L, Kaati L, Mao W, Wang GA (eds) Proceedings of the 2016 IEEE Conference on Intelligence and Security Informatics. The Printing House, Stoughton, WI, USA, pp 19–24. https://doi.org/10.1109/ISI.2016.7745437
13. Allodi L (2017) Economic factors of vulnerability trade and exploitation. In: Thuraisingham B, Evans D, Malkin T, Xu D (eds) Proceedings of the 2017 ACM SIGSAC Conference on

Computer and Communications Security. ACM, New York, pp 1483–1499. https://doi.org/10.1145/3133956.3133960
14. Bullough BL, Yanchenko AK, Smith CL, Zipkin JR (2017) Predicting exploitation of disclosed software vulnerabilities using open-source data. In: Verma R, Thuraisingham B (eds) Proceedings of the 3rd ACM on International Workshop on Security and Privacy Analytics. ACM, New York, pp 45–53. https://doi.org/10.1145/3041008.3041009
15. Allodi L, Shim W, Massacci F (2013) Quantitative assessment of risk reduction with cybercrime black market monitoring. In: 2013 IEEE Security and Privacy Workshops. IEEE Computer Society, Los Alamitos, CA, USA, pp 165–172. https://doi.org/10.1109/SPW.2013.16
16. Bozorgi M, Saul LK, Savage S, Voelker GM (2010) Beyond heuristics: learning to classify vulnerabilities and predict exploits. In: Rao B, Krishnapuram B, Tomkins A, Yang Q (eds) Proceedings of the 16th ACM SIGKDD International Conference on Knowledge Discovery and Data Mining. ACM, New York, pp 105–114. https://doi.org/10.1145/1835804.1835821
17. Motoyama M, McCoy D, Levchenko K, Savage S, Voelker GM (2011) An analysis of underground forums. In: Thiran P, Willinger W (eds) Proceedings of the 2011 ACM SIGCOMM Conference on Internet Measurement. ACM, New York, pp 71–80. https://doi.org/10.1145/2068816.2068824
18. Holt TJ, Lampke E (2010) Exploring stolen data markets online: products and market forces. Crim Justice Stud 23(1):33–50. https://doi.org/10.1080/14786011003634415
19. Shakarian J, Gunn AT, Shakarian P (2016) Exploring malicious hacker forums. In: Jajodia S, Subrahmanian V, Swarup V, Wang C (eds) Cyber deception. Springer, Cham, pp 259–282. https://doi.org/10.1007/978-3-319-32699-3_11
20. Nunes E, Diab A, Gunn A, Marin E, Mishra V, Paliath V, Robertson J, Shakarian J, Thart A, Shakarian P (2016) Darknet and deepnet mining for proactive cybersecurity threat intelligence. In: Chen H, Hariri S, Thuraisingham B, Zeng D (eds) Proceedings of the 2016 IEEE Conference on Intelligence and Security Informatics, pp 7–12. https://doi.org/10.1109/ISI.2016.7745435
21. Robertson J, Diab A, Marin E, Nunes E, Paliath V, Shakarian J, Shakarian P (2017) Darkweb cyber threat intelligence mining. Cambridge University Press, New York. https://doi.org/10.1017/9781316888513
22. Liu Y, Sarabi A, Zhang J, Naghizadeh P, Karir M, Bailey M, Liu M (2015) Cloudy with a chance of breach: forecasting cyber security incidents. In: Proceedings of the 24th USENIX Security Symposium. USENIX Association, Berkeley, CA, USA, pp 1009–1024. https://www.usenix.org/sites/default/files/sec15_full_proceedings.pdf
23. Soska N, Christin K (2014) Automatically detecting vulnerable websites before they turn malicious. In: Proceedings of the 23rd USENIX Security Symposium. USENIX Association, Berkeley, CA, USA, pp 625–640. https://www.usenix.org/sites/default/files/sec14_full_proceedings.pdf
24. Almukaynizi M, Nunes E, Dharaiya K, Senguttuvan M, Shakarian J, Shakarian P (2017) Proactive identification of exploits in the wild through vulnerability mentions online. In: Sobiesk E, Bennett D, Maxwell P (eds) Proceedings of the 2017 International Conference on Cyber Conflict. Curran Associates, Red Hook, NY, USA, pp 82–88. https://doi.org/10.1109/CYCONUS.2017.8167501
25. Zhang S, Caragea D, Ou X (2011) An empirical study on using the national vulnerability database to predict software vulnerabilities. In: Hameurlain A, Liddle SW, Schewe KD, Zhou X (eds) Database and expert systems applications. Springer, Heidelberg, pp 217–231. https://doi.org/10.1007/978-3-642-23088-2_15
26. Hao S, Kantchelian A, Miller B, Paxson V, Feamster N (2016) PREDATOR: proactive recognition and elimination of domain abuse at time-of-registration. In: Weippl E, Katzenbeisser S, Kruegel C, Myers A, Halevi S (eds) Proceedings of the 2016 ACM SIGSAC Conference on Computer and Communications Security. ACM, New York, pp 1568-1579. https://doi.org/10.1145/2976749.2978317
27. Cortes C, Vapnik V (1995) Support-vector networks. Mach Learn 20(3):273–297. https://doi.org/10.1023/A:1022627411411

28. Chawla NV, Bowyer KW, Hall LO, Kegelmeyer WP (2002) SMOTE: synthetic minority over-sampling technique. J Artif Int Res 16(1):321–357. https://doi.org/10.1613/jair.953
29. Allodi L, Massacci F, Williams JM (2017) The work-averse cyber attacker model: theory and evidence from two million attack signatures. https://doi.org/10.2139/ssrn.2862299
30. Breiman L (2001) Random forests. Mach Learn 45(1):5–32. https://doi.org/10.1023/A:1010933404324
31. Breiman L (1996) Bagging predictors. Mach Learn 24(2):123–140. https://doi.org/10.1007/BF00058655
32. Pedregosa F, Varoquaux G, Gramfort A, Michel V, Thirion B, Grisel O, Blondel M, Prettenhofer P, Weiss R, Dubourg V, Vanderplas J, Passos A, Cournapeau D, Brucher M, Perrot M, Duchesnay É (2011) Scikit-learn: machine learning in Python. J Mach Learn Res 12:2825–2830
33. Guo D, Shamai S, Verdu S (2005) Mutual information and minimum mean-square error in Gaussian channels. IEEE Trans Inform Theory 51(4):1261–1282. https://doi.org/10.1109/TIT.2005.844072
34. Galar M, Fernandez A, Barrenechea E, Bustince H, Herrera F (2012) A review on ensembles for the class imbalance problem: bagging-, boosting-, and hybrid-based approach. IEEE Trans Syst Man Cybern C 42(4):463–484. https://doi.org/10.1109/TSMCC.2011.2161285
35. Barreno M, Bartlett PL, Chi FJ, Joseph AD, Nelson B, Rubinstein BIP, Saini U, Tygar JD (2008) Open problems in the security of learning. In: Balfanz D, Staddon J (eds) Proceedings of the 1st ACM Workshop on AISec. ACM, New York, pp 19–26. https://doi.org/10.1145/1456377.1456382
36. Barreno M, Nelson B, Joseph AD, Tygar J (2010) The security of machine learning. Mach Learn 81(2):121–148. https://doi.org/10.1007/s10994-010-5188-5
37. Biggio B, Nelson B, Laskov P (2011) Support vector machines under adversarial label noise. In: Hsu C-N, Lee WS (eds) Proceedings of the 3rd Asian Conference on Machine Learning, pp 97–112. http://www.jmlr.org/proceedings/papers/v20/biggio11/biggio11.pdf

Chapter 5
Applying Artificial Intelligence Methods to Network Attack Detection

Alexander Branitskiy and Igor Kotenko

Abstract This chapter reveals the methods of artificial intelligence and their application for detecting network attacks. Particular attention is paid to the representation of models based on neural, fuzzy, and evolutionary computations. The main object is a binary classifier, which is designed to match each input object to one of two sets of classes. Various schemes for combining binary classifiers are considered, which allows building models trained on different subsamples. Several optimizing techniques are proposed, both in terms of parallelization (for increasing the speed of training) and usage of aggregating compositions (for enhancing the classification accuracy). Principal component analysis is also considered, which is aimed at reducing the dimensionality of the analyzed attack feature vectors. A sliding window method was developed and adopted to decrease the number of false positives. Finally, the model efficiency indicators obtained during the experiments using the multifold cross-validation are provided.

5.1 Introduction

Network attack detection is a difficult task, and there are lots of methods for it in the modern literature. Some of these are based on such rapidly developing fields as artificial intelligence. These include neural networks, fuzzy logic, and genetic algorithms.

The construction of attack detection systems plays one of the key roles in ensuring the security of network nodes. Therefore it is important to use advanced tools, including artificial intelligence methods, and apply them to network attack detection problems.

A. Branitskiy · I. Kotenko (✉)
St. Petersburg Institute for Informatics and Automation of the Russian Academy of Sciences, St. Petersburg, Russia
e-mail: ivkote@comsec.spb.ru

© Springer Nature Switzerland AG 2019

L. F. Sikos (ed.), *AI in Cybersecurity*, Intelligent Systems Reference Library 151,
https://doi.org/10.1007/978-3-319-98842-9_5

From the input data interpretation point of view, the methods of network attack detection are divided into anomaly detection methods and misuse detection methods [1].

In the case of anomaly detection methods, the system should contain data on the normal behavior of the analyzed object (user, process, sequence of network packets)—a normal behavior pattern. If there is a discrepancy between the observed parameters and the parameters of the normal behavior pattern, the system detects a network anomaly. Adding and modifying of such patterns can be performed both in manual mode and automatically; and some parameters of the normal behavior pattern may vary depending on the time of day. The drawback of anomaly detection systems is the presence of false positives, which can be caused by incompleteness of the available set of normal behavior patterns. In contrast, misuse detection methods identify only those illegal actions for which there is an exact representation in the form of attack templates. By attack pattern we mean some set of rules, e.g., pattern matching or signature search, which explicitly describes a specific attack. Applying such rules to the fields of the identified object gives an unambiguous answer about its belonging to this attack. The main problem of designing misuse detection systems is how to provide a mechanism for fast search based on specified signature rules. The approach discussed in this chapter is based on artificial intelligence methods, and uses parameters of normal and anomalous behavior as training data for binary classifiers.

The chapter is organized as follows. In Sect. 5.2, we consider some papers that concern the issue of network attack detection using hybrid approaches. Section 5.3 describes the models of binary classifiers and algorithms for their training. In Sect. 5.4, we introduce a sliding window method for calculating the network parameters, a genetic optimization of binary classifier weights, and an algorithm for network attack detection using a binary classifier. In Sect. 5.5, we propose to consider several combining schemes [2–4], which allows to construct multi-class models based on binary classifiers. The results of the experiments are presented in Sect. 5.6.

5.2 Related Work

In this section we present some papers devoted to the investigated problem.

In [5, 6] three classifiers were considered: decision tree, support vector machine (SVM), and a combination of these, where the hybrid classifier consisted of two phases. First, the test data were passed to decision trees, which generated a node information as numbers of leaves. Then the test data with node information were processed by the SVM, which output is a classification result. It is suggested that while using this approach, the additional information obtained from the decision tree will allow to enhance the effectiveness of the SVM. If these three classifiers give different results, the ultimate result is based on weight voting.

In [7] the set of three neural networks and SMV is considered. The output value of the hybrid classifier is a weighted sum of output values of these classifiers. The weights are calculated on the basis of the *mean square error* (*MSE*) value.

In [8] the authors propose to use the output values of neural networks as the input values for the procedure of weight voting and majority voting. Using the test sample consisting of 6,890 instances, the classification rate was achieved above 99%.

In [9] the two-level scheme of network attack detection is described. To this end, several adaptive neuro-fuzzy classifiers are combined together. Each of these classifiers is designed for detecting only one attack. The final classification is performed by a fuzzy module, which implements a system of Mamdani fuzzy inference with two membership functions. The module assignment works by determining how anomalous the network record is; the class of this record is a class of the first-level fuzzy module with the greatest output value.

For solving the task of attack detection, in [10] it was proposed to use K radial basis neural networks. Each of these networks is trained on different disjoint subsets $D_1, \ldots, D_K$ of the original training dataset D. Such subsets are generated using the method of fuzzy clustering, according to which each element $\vec{z} \in D$ belongs to the region D_i with a certain membership degree of $u_i^{\vec{z}}$. Each subset D_i $(i = 1, \ldots, K)$ is composed of those elements that have the greatest membership degree with respect to this subset among all other subsets. As the authors emphasize, due to such preliminary decomposition, the generalization ability of base classifiers is improved and time of their training is reduced, because only those objects are employed for their configuration which are most densely grouped around the formed center of the training subset. To combine the output results $\vec{y_1}, \ldots, \vec{y_K}$ of these classifiers, taking the vector $\vec{z}$ as an input argument, the multilayer neural network is used. Its input vector is represented as a set of elements obtained by applying a threshold function to each component of the vector $u_i^{\vec{z}} \cdot \vec{y_i}$ $(i = 1, \ldots, K)$. A similar approach has been used previously in [11], where feed-forward neural networks served as base classifiers and the input for an aggregating module was composed of the direct values of vectors $u_1^{\vec{z}} \cdot \vec{y_1}, \ldots, u_k^{\vec{z}} \cdot \vec{y_K}$.

In [12], the analysis of the records of network connections was performed by neuro-fuzzy models and SVM. The authors determined four main stages in the proposed approach. In the first stage, the generation of training data is performed by a K-means clustering method. The second stage was training of neuro-fuzzy classifiers. In the third stage, the input vector was formed for SVM. The final stage was attack detection using the latter classifier.

In [13], an individual neural network with a single *hidden layer*[1] was constructed for detecting each of the three types of DDoS attacks, which were conducted with the use of protocols TCP, UDP, and ICMP. The last layer of each of these neural networks consists of a single node, the output value of which is interpreted as the presence or absence of DDoS attacks of the appropriate type. The proposed approach

[1]A hidden layer is a layer hidden between the input and output layers, whose output is the input of another layer. Neural networks with more than one hidden layer are called deep neural networks, in which machine learning is called deep learning.

was implemented as a module in the Snort, and tested on real traffic network environments.

In [14], to detect DoS attacks, it was proposed to use an approach, which combines the method of normalized entropy for calculating the feature vectors and SVM for their analysis. In order to detect anomalies, six parameters were extracted from network traffic, numerically expressed as the occurrence intensity of different values of selected fields within the packets during the 60 second window. In this approach, the network parameters are calculated using the method of normalized entropy, and they are used as the input for training and testing data based on SVM.

In [15], to detect DoS attacks and scanning hosts, the authors considered an approach based on consecutive application of a vector compression procedure and two fuzzy transformations. First, principal component analysis is applied to the input eight-dimensional feature vector of network connections, which allows to reduce its dimension to five components while preserving the relative total variance at a level of more than 90%. The next step is the training or testing of the neuro-fuzzy network, the output value of which is processed by the method of fuzzy clustering.

Our approach is based on these papers, but we suggest to apply the binary classifiers for constructing the multi-class model for recognizing different types of attacks. This approach is more flexible, which allows to construct a variety of combining schemes without strict binding to the aggregating compositions.

5.3 Binary Classifiers

The following sections discuss popular models for binary classification.

5.3.1 Neural Networks

This section describes neural networks in general and some training algorithms. By structure and function, an artificial *neural network* is an analogue to the human brain with computing nodes corresponding to neurons, and relations corresponding to synapses. The simulation of perturbing nervous impulse transmission from one neuron to another is reduced to establish corresponding connections between the elements of the considered computational structure. The presence and strength of relations between neurons can be set by specifying non-zero weight coefficients whose values proportionally express the significance of input signals. The stronger the input signal, the more multiplicative weight is assigned to the corresponding connection. After configuring the similar structures described by at least two layers, they can perform a sufficiently accurate approximation of the training instances [16–19].

A biological neuron consists of a nucleus, a soma (cell body), and multiple appendages [20]. Appendages of the first type, known as axons, are represented in a single copy in each neuron and serves as an original transmitter of the nervous impulse generated by this neuron. Appendages of the second type, known as dendrites, receive the signals coming from the axons of adjacent cells. A region of nerve fibers located at the junction of the axon and dendrite is called a synapse. Depending on the type, this element can cause agitation or inhibition of the transmitted signals that turns accordingly into amplification or attenuation of relations between neurons when projected onto an artificial analogue of a neuron.

The input layer of a neural network is a dummy layer that performs the predistribution of input signals before they are treated. The input vector for each node of the this layer[2] is the scalar product of the synaptic weights vector and the input vector $\vec{z} = (z_1, \ldots, z_n)^T$. The signal on the input of the i'th neuron of the first hidden layer consisting of N_1 nodes is constructed as follows: $z_{i'}^{(1)} = \sum_{j=1}^{n} w_{i'j}^{(1)} \cdot z_j + \theta_{i'}^{(1)}$, where $i' = 1, \ldots, N_1$, $\left\{w_{i'j}^{(1)}\right\}_{j=1}^{n}$, are weights, specifying a conversion of the signal $\vec{z}$ on the input of the i'th neuron of the first hidden layer, and $\theta_{i'}^{(1)}$ is a bias of the i'th neuron, located in the first hidden layer. The output of this neuron can be considered as value $y_{i'}^{(1)} = \varphi\left(z_{i'}^{(1)}\right)$. Similarly, the input and output signals are set for each i''th neuron located in the second hidden layer with N_2 neurons: $z_{i''}^{(2)} = \sum_{j=1}^{N_1} w_{i''j}^{(2)} \cdot y_j^{(1)} + \theta_{i''}^{(2)}$ and $y_{i''}^{(2)} = \varphi\left(z_{i''}^{(2)}\right)$, where $i'' = 1, \ldots, N_2$, $\left\{w_{i''j}^{(2)}\right\}_{j=1}^{N_1}$ are weights, specifying a conversion of the signals $\vec{y^{(1)}} = (y_1^{(1)}, \ldots, y_{N_1}^{(1)})^T$ on the input of the i''th neuron, which is located in the second hidden layer, $\theta_{i''}^{(2)}$ is a bias of the i''th neuron, and φ is an activation function. The resulting signal $y_1^{(3)}$ is constructed as follows: $y_1^{(3)} = \varphi\left(\sum_{j=1}^{N_2} w_{1j}^{(3)} \cdot y_j^{(2)} + \theta_1^{(3)}\right)$, where $\left\{w_{1j}^{(3)}\right\}_{j=1}^{N_2}$ are weights on the input of the last layer neuron, and $\theta_1^{(3)}$ is a bias of thc output neuron (see Fig. 5.1).

Generally speaking, the following formula can be used to describe how the binary neural network model functions:

$$Y(\vec{z}) = \varphi\left(\sum_{i=1}^{N_2} w_{1i}^{(3)} \cdot \varphi\left(\sum_{j=1}^{N_1} w_{ij}^{(2)} \cdot \varphi\left(\sum_{k=1}^{n} w_{jk}^{(1)} \cdot z_k + \theta_j^{(1)}\right) + \theta_i^{(2)}\right) + \theta_1^{(3)}\right).$$

The most common learning algorithm of multilayer neural networks is the backpropagation[3] algorithm (see Algorithm 3).

[2]The first hidden layer is intended to perform sequential operations of linear and nonlinear transformations.

[3]Backpropagation is a method for calculating a gradient needed in weight calculations.

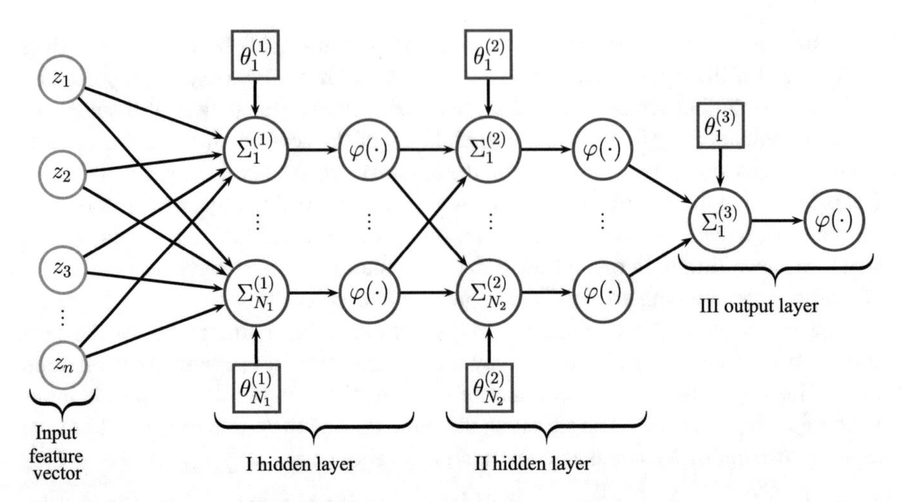

Fig. 5.1 Three-layer neural network

Algorithm 3: Algorithm for training multilayer neural networks

---

1. Set the neural network structure, i.e., choose activation function types and the number of hidden layers and neurons located there.
2. Specify the maximum number of training epochs T and the minimum value of the total mean square error ε.
3. Set the counter of current iterations to zero, i.e., $t := 0$, and initialize the weight coefficients $w_{ij}^{(K)}$ by arbitrary values, where K denotes a layer number, i corresponds to a neuron position number in the Kth layer, j indicates a connection between the current neuron and the output signal of the jth neuron in the (K–1)th layer.
4. Perform steps 4a–4c for each vector $\vec{x_k}$ $(k = 1, \ldots, M)$.

   a. Perform a feed-forward propagation of the signals—calculate input signals for each ith neuron in the Kth layer according to the formula $x_i^{(K)} = \sum_{j=1}^{N_{K-1}+1} w_{ij}^{(K)} \cdot y_j^{(K-1)}$, where N_{K-1} is the number of neurons in the $(K-1)$th layer, $w_{ij}^{(K)} = \theta_i^{(K)}$ and $y_j^{(K-1)} = 1$ for $j = N_{K-1} + 1$, $y_j^{(K-1)} = \varphi\left(x_j^{(K-1)}\right)$ for others $K > 1$ and $y_j^{(K-1)} = x_{k_j}$ (original signal) for $K = 1$.
   b. Perform backpropagation of the error: calculate an augmentation of weight coefficients of neurons by the formula $\Delta w_{ij}^{(K)} = \alpha \cdot \delta_i^{(K)} \cdot y_j^{(K-1)}$, successively starting with the last and ending with the first layer, where α is the correction proportionality coefficient of weights and $0 < \alpha \leqslant 1$. If the Kth layer is output, then $\delta_i^{(K)} = \varphi'\left(x_i^{(K)}\right) \cdot \left(u_{k_i} - y_i^{(K)}\right)$, otherwise $\delta_i^{(K)} = \varphi'\left(x_i^{(K)}\right) \cdot \sum_{j=1}^{N_{K+1}+1} \delta_j^{(K+1)} \cdot w_{ji}^{(K+1)}$, where u_{ki} denotes the desired

output of the neural network in the ith neuron of the output layer for the kth training vector.

c. Adjust the weight coefficients of neurons by the formula $w_{i_j}^{(K)} := w_{i_j}^{(K)} + \Delta w_{i_j}^{(K)}$.

5. Increase the counter of current iterations, i.e., $t := t + 1$.
6. Terminate the algorithm if one of the conditions is satisfied, formally $t \geqslant T$ or $\sum_{k=1}^{M} E(\vec{x_k}) \leqslant \varepsilon$, where $E(\vec{x_k}) = \frac{1}{2} \cdot \sum_{i=1}^{N_{K_{all}}} \left(u_{k_i} - y_i^{(K_{all})}\right)^2$ is the total mean square error of the neural network, which has the output value $y^{(\vec{K}_{all})} = \left(y_1^{(K_{all})}, \ldots, y_{N_{K_{all}}}^{(K_{all})}\right)^T$ and consists of three layers and a neuron on the output layer ($K_{all} = 3$, $N_{K_{all}} = 1$) if vector $\vec{x_k}$ is passed to the distributed layer (with the desired output vector $\vec{u_k} = \left(u_{k1}, \ldots, u_{k N_{K_{all}}}\right)^T$); otherwise jump to step 4.

The above algorithm belongs to the general family of gradient descent algorithms in which a search of the minimum point is performed in the direction opposite to the gradient of the function to be optimized (e.g., MSE). Falling into a local minimum pit is a characteristic of such algorithms, meaning that these algorithms can barely modify the weight parameters, despite the presence of a deeper extremum compared to what has already been found. These issues are solved partially by means of various improvements of the backpropagation algorithm [20]. First a feed-forward propagation of the signals is performed, then an adjustmentment is calculated for each weight. Some of such modifications use a variable correction proportionality coefficient of weights, which depends on maintaining or changing the derivative sign [21], taking into account the momentum factors for changing each individual weight [22], or considering the second derivatives [23, 24].

Radial basis function networks are constructed differently than multilayer neural networks. The first hidden layer of such networks is designed for the projection of the input vector $\vec{z}$ into the new feature space[4] $\left(\Phi\left(\vec{z} - \vec{w_1^{(1)}}\right), \ldots, \Phi\left(\vec{z} - \vec{w_{N_1}^{(1)}}\right)\right)^T$. In such a space, the original n-dimensional vector is transformed into an N_1-dimensional vector. Each component of a new vector reflects the degree of proximity of input vector $\vec{z} = (z_1, \ldots, z_n)^T$ and weight vector $\vec{w_i^{(1)}} = \left(w_{i_1}^{(1)}, \ldots, w_{i_n}^{(1)}\right)^T$, which is assigned to the ith neuron of the first hidden layer, where $i = 1, \ldots, N_1$. Here $\Phi\left(\vec{z} - \vec{w_i^{(1)}}\right) = \exp\left\{-\frac{\|\vec{z} - \vec{w_i^{(1)}}\|^2}{s}\right\} = \exp\left\{-\frac{\sum_{j=1}^{n}\left(z_j - w_{i_j}^{(1)}\right)^2}{s}\right\}$ is a radial basis function. A very unique feature is that the weights of this layer are not changed during the training process; instead they are set static using one of the following approaches:

---

[4] In machine learning, a feature space is a vector spacc associated with feature vectors, i.e., n-dimensional vectors of numerical features that represent objects.

(1) random initialization, (2) initialization by randomly selected training vectors, (3) initialization using the clustering method. In the experiments, the third approach was used, namely the K-means method, in which the number of clusters was equal to the dimension of the hidden layer, N_1. The method is aimed at constructing points $\left\{\vec{\bar{v}}_i\right\}_{i=1}^{K}$, known as centroids, and arranged in such a way that the summary distance from them to the points, located in one of the vicinities of points $\left\{\vec{\bar{v}}_i\right\}_{i=1}^{K}$, is minimal.

The model of the radial basis function network is represented as follows:

$$Y(\vec{z}) = \varphi\left(\sum_{i=1}^{N_1} w_{1_i}^{(2)} \cdot \Phi\left(\vec{z} - \vec{w}_i^{(1)}\right) + \theta_1^{(2)}\right).$$

The algorithm for training the radial basis function network is given in Algorithm 4.

Algorithm 4: Algorithm for training the radial basis function network

1. Set a neural network structure, i.e., choose the dimension of the hidden layer N_1 and activation function type φ on thc output layer.
2. Specify the maximum number of training epochs T and the minimum value of the total MSE ε.
3. Calculate the centroids $\left\{\vec{\bar{v}}_i\right\}_{i=1}^{N_1}$ of clusters using the K-means method and initialize the weight coefficients $w_{i_j}^{(1)}$ of the neurons on the first hidden layer by the components of these centroids, i.e., $w_{i_j}^{(1)} := \bar{v}_{i_j}$ $(i = 1, \ldots, N_1, j = 1, \ldots, n)$.
4. Set the counter of current iterations to zero $(t := 0)$, and initialize the weight coefficients $w_{i_j}^{(2)}$ of the neurons on the second output layer by arbitrary values $(i = 1, \ldots, N_2, j = 1, \ldots, N_1)$.
5. Perform steps 5a–5b for each vector $\vec{x}_k$ $(k = 1, \ldots, M)$.

   a. Perform a feed-forward propagation of the signals—calculate output signals for each ith neuron located in the second layer according to the formula $y_i^{(2)} = \varphi\left(x_i^{(2)}\right)$, where $x_i^{(2)} = \sum_{j=1}^{N_1+1} w_{ij}^{(2)} \cdot y_j^{(1)}$; $w_{ij}^{(2)} = \theta_i^{(2)}$ and $y_j^{(1)} = 1$ for $j = N_1 + 1$; $y_j^{(1)} = \Phi\left(\vec{x}_k - \vec{w}_j^{(1)}\right)$ for $j = 1, \ldots, N_1$.
   b. Update the weight coefficients of the output layer of the neural network by the Widrow-Hoff rule $w_{ij}^{(2)} := w_{ij}^{(2)} + \Delta w_{ij}^{(2)}$ [20], where $\Delta w_{ij}^{(2)} = \alpha \cdot \varphi'\left(x_i^{(2)}\right) \cdot \left(u_{ki} - y_i^{(2)}\right) \cdot y_j^{(1)}$, u_{ki} is the desired output of the neural network in the ith neuron on the output layer for the kth instance from the training sample.

6. Increase the counter of current iterations, i.e., $t := t + 1$.

7. Terminate the algorithm if one of the conditions is satisfied, formally $t \geqslant T$ or $\sum_{k=1}^{M} E(\vec{x_k}) \leqslant \varepsilon$, where $E(\vec{x_k}) = \frac{1}{2} \cdot \sum_{i=1}^{N_{K_{all}}} \left(u_{ki} - y_i^{(K_{all})}\right)^2$—a total MSE of the neural network, which has the output value $y^{(\vec{K}_{all})} = \left(y_1^{(K_{all})}, \ldots, y_{N_{K_{all}}}^{(K_{all})}\right)^T$ and consists of 2 layers and 1 neuron on the output layer ($K_{all} = 2, N_{K_{all}} = 1$), if the vector $\vec{x_k}$ is passed to the distributed layer (with the desired output vector $\vec{u_k} = \left(u_{k1}, \ldots, u_{kN_{K_{all}}}\right)^T$); otherwise jump to step 5.

The model of the Jordan recurrent network is an extension of the multilayer neural network model described above [25], which introduces an input-output feedback into the context layer as follows:

$$Y(\vec{z}) = \varphi\left(\sum_{i=1}^{N_1} w_{1i}^{(2)} \cdot \varphi\left(\sum_{j=1}^{n} w_{ij}^{(1)} \cdot z_j + w_{i0}^{(1)} \cdot z_0 + \theta_i^{(1)}\right) + \theta_1^{(2)}\right).$$

The output signal z_0 resulting from the analysis of one of the previous vectors is stored for several cycles before processing the new vector complemented by the value of this signal. This allows to remember some history of alternation of images of anomalous and normal network connections. In principle, the rest of the functioning of such networks is similar to the training of multilayer neural networks (see Algorithm 5).

Algorithm 5: Algorithm for training the recurrent neural network

1. Initialize the values, i.e., $z_0 := 0$.
2. Perform steps 1–6 of the algorithm for training the multilayer neural network and store the output value in variable z_0 after each iteration.

5.3.2 *Neuro-Fuzzy Networks*

In this section, we discuss neuro-fuzzy networks and an algorithm for their training.

An approach used in the construction of intelligent kernel for detecting the network anomalies is *neuro-fuzzy networks* [26]. This approach reflects the ability of the human mind to make decisions under conditions of uncertainty and ambiguity.

Typically, such systems consist of five functional blocks [27]. The first block is a rule database, which includes a set of fuzzy implications (rules) in the form "if *A* then *B*," where *A* is called the premise (antecedent), and *B* is called the consequence (succedent). Such rules are mainly different from the traditional production rules in a way that each of the statements included in *A* and *B* is assigned (attributed to) a certain number between 0 and 1 to indicate the confidence degree of the premise and the consequence. The second block is a database that contains a set of membership functions. Such functions set the input linguistic variables to the transition from their crisp values to fuzzy linguistic terms. For each of such terms, an individual membership function is constructed, whose output value characterizes the measure of matching the input variable to a corresponding fuzzy set (term). The most frequently used membership functions are continuous piecewise differentiable (triangular and trapezoidal) functions or smooth functions (family of bell-shaped functions) with the range [0, 1] or (0, 1]. The third block is a fuzzification block whose role is to apply the specified membership function to the input argument. Each of the conjuncts A_i, which is a part of premise $A = A_1 \wedge \ldots \wedge A_n$, and consequence B are represented as the fuzzy statements z_i is γ_i and y is Γ, respectively, where z_i and y are the linguistic variables, and γ_i and Γ are the linguistic terms ($i = 1, \ldots, n$). The result of a fuzzification process is the calculated values of these fuzzy statements. The fourth block is a fuzzy inference block, which contains a set of fuzzy implications already integrated into its kernel and provides a mechanism (e.g., rules modus ponens or modus tollens) for calculating consequence B using the input set of conjuncts in premise A. To calculate the full verity degree of the left-hand side, T-norms are used, the most common examples of which are the minimum and product operators. On the output of the fuzzy inference block, one or more fuzzy terms along with the corresponding values membership functions are formed for linguistic variable y. The fifth block is the defuzzification block, which restores the quantitative value of the linguistic variable y by its fuzzy values. Namely, the data obtained through the fuzzy inference block operation is converted into quantitative values using one of the following methods: center of area, centre of gravity, center of sums, or the maximum of the membership function.

In the above fuzzy inference system, consequence B in all rules "if A then B" had a type of the fuzzy statement y is Γ, which is not dependent on the linguistic variables encountered as a part of premise A. The approach proposed by Takagi and Sugeno is aimed at eliminating this drawback, and is based on introducing into the right-hand side of each rule a certain functional dependence on elements of its left-hand side, i.e., $y = f(z_1, \ldots, z_n)$ [26]. In real-world situations, one often has to deal with models of this type, in particular, when a person or device is not able to estimate accurately the values of the input parameters, but at the same time the control effect can be explicitly calculated by the known formula.

Neuro-fuzzy networks, also known as adaptive neuro-fuzzy inference systems (ANFIS) [27], are extensions of the Takagi-Sugeno model, which possess an element of adaptive adjustment (training) of parameters. These networks consist of five layers, where the input signals undergo changes, propagating sequentially from the first to the last layer. Each fuzzy rule in these networks is represented as an element belonging to a set of rules of the form

$$\left\{\text{if } \left(z_1 \text{ is } \gamma_1^{(j_1)} \wedge \ldots \wedge z_n \text{ is } \gamma_n^{(j_n)}\right) \text{ then} \right.$$
$$\left. y = f^{(j)}(z_1, \ldots, z_n) = p_0^{(j)} + p_1^{(j)} \cdot z_1 + \cdots + p_n^{(j)} \cdot z_n \right\}_{j=1}^{Q}.$$

Here Q denotes a cardinality of the fuzzy rule set, in which each variable $z_1, \ldots, z_n$ has exactly r fuzzy terms; $j_1, \ldots, j_n$ denote numbers of fuzzy terms, corresponding to linguistic variables $z_1, \ldots, z_n$, in the fuzzy rule at number j ($1 \leqslant j_1 \leqslant r, \ldots, 1 \leqslant j_n \leqslant r$). As in classical fuzzy inference systems, the left-hand side of this rule is the conjunction of fuzzy statements, which expresses the degree of matching the input crisp value z_i to a particular linguistic term $\gamma_i^{(j_i)}$ according to value $\mu_{\gamma_i^{(j_i)}}(z_i)$, where the most frequently used membership functions $\mu_{\gamma_i^{(j_i)}}(z_i)$ are the bell-shaped function $\mu_{\gamma_i^{(j_i)}}(z_i) = \left(1 + \left|{}^{(z_i - c_{ij})}/_{a_{ij}}\right|^{2 \cdot b_{ij}}\right)^{-1}$ or the Gaussian function $\mu_{\gamma_i^{(j_i)}}(z_i) = \exp\left\{-\left({}^{(z_i - c_{ij})}/_{a_{ij}}\right)^2\right\}$, where $i = 1, \ldots, n$ and $j = 1, \ldots, Q$.

In Fig. 5.2, a neuro-fuzzy network is shown.

The node elements of the first layer in a neuro-fuzzy network act as a fuzzification of the input linguistic variable $z_{i'}$, and the output vector of this layer are the values

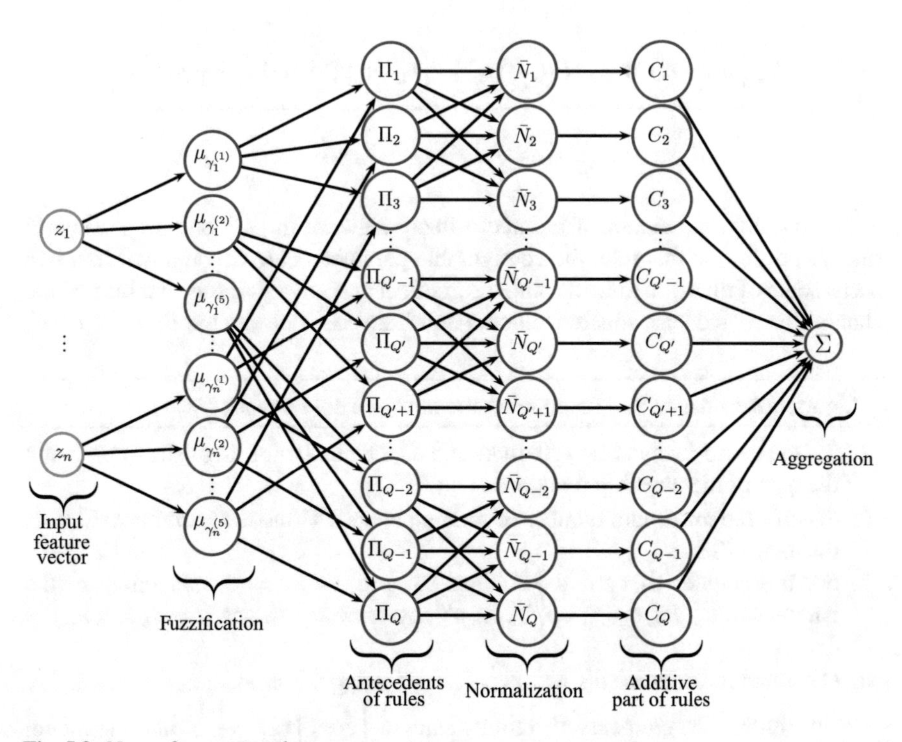

Fig. 5.2 Neuro-fuzzy network

of membership function $\mu_{\gamma_{i'}^{(j')}}$ of this variable to the j'th fuzzy set (term) $\gamma_{i'}^{(j')}$: $Y_{j'i'}^{(1)} = \mu_{\gamma_{i'}^{(j')}}(z_{i'})$, where $i' = 1, \ldots, n, j' = 1, \ldots, r$. The second layer performs the creation of premises of fuzzy rules with their association with using the T-norm operation (product); the kth output value of this layer can be considered as the weight assigned to the kth rule $Y_k^{(2)} = Y_{k_1 1}^{(1)} \times \ldots \times Y_{k_n n}^{(1)} = \mu_{\gamma_1^{(k_1)}}(z_1) \times \ldots \times \mu_{\gamma_n^{(k_n)}}(z_n)$, where $k = 1, \ldots, Q$. Let us emphasize that $Q \leqslant r^n$ in the absence of any conflicting rules. The ratio of a rule weight and the total sum of the weights of all the rules is calculated in the elements of the third layer; the output of this layer is the value normalized to [0, 1]: $Y_k^{(3)} = \frac{Y_k^{(2)}}{\sum_{i=1}^{Q} Y_i^{(2)}}$. The result of the consequence of each of the rules, taking into account the relative degree of its fulfillment received on the third layer, is calculated on the fourth layer. The kth output value of this layer represents an additive portion of the kth rule in the entire output of the network, i.e., $Y_k^{(4)} = Y_k^{(3)} \cdot f^{(k)}(z_1, \ldots, z_n) = Y_k^{(3)} \cdot \left(p_0^{(k)} + p_1^{(k)} \cdot z_1 + \cdots + p_n^{(k)} \cdot z_n\right)$. At the fifth output layer, there is a single neuron responsible for summing the input signals received from the fourth layer nodes, formally $Y^{(5)} = \sum_{i=1}^{Q} Y_i^{(4)} = \sum_{i=1}^{Q} Y_i^{(3)} \cdot f^{(i)}(z_1, \ldots, z_n) = \frac{\sum_{i=1}^{Q} Y_i^{(2)} \cdot f^{(i)}(z_1, \ldots, z_n)}{\sum_{i=1}^{Q} Y_i^{(2)}}$.

Thereby the ANFIS model is represented as $Y(\vec{z}) =$

$$= \frac{\sum\limits_{i=1}^{Q} \left(\mu_{\gamma_1^{(i_1)}}(z_1) \times \ldots \times \mu_{\gamma_n^{(i_n)}}(z_n)\right) \cdot \left(p_0^{(i)} + p_1^{(i)} \cdot z_1 + \cdots + p_n^{(i)} \cdot z_n\right)}{\sum\limits_{i=1}^{Q} \mu_{\gamma_1^{(i_1)}}(z_1) \times \ldots \times \mu_{\gamma_n^{(i_n)}}(z_n)}.$$

For training, the author of the neuro-fuzzy network in [27] proposed a hybrid rule. According to this rule, the configurable parameters are decomposed into two parts: some of them are trained using the gradient descent algorithm, and the rest are changed using the least square method. Algorithm 6 is such an algorithm.

Algorithm 6: Algorithm for training the neuro-fuzzy network

---

1. Set the neuro-fuzzy network structure, i.e., the number of linguistic terms r and the type of membership functions $\mu_{\gamma_i^{(j)}}$;
2. Specify the maximum number of training epochs T and the minimum value of the total MSE ε;
3. Set the counter of current iterations to zero, i.e., $t := 0$, and initialize the parameters $a_{ij}, b_{ij}, c_{ij}, p_0^{(j)}, p_1^{(j)}, \ldots, p_n^{(j)}$ by arbitrary values ($i = 1, \ldots, n, j = 1, \ldots, Q$).
4. Calculate the coefficients $p_0^{(j)}, p_1^{(j)}, \ldots, p_n^{(j)}$ using the least square method. By substituting the elements of training sample $\left\{\vec{x}_i = \left\{x_{ij}\right\}_{j=1}^{n}\right\}_{i=1}^{M}$ into the model formula, the following system of equations can be obtained:

$$\underbrace{\begin{pmatrix} e_{11} & e_{11}x_{11} & \ldots & e_{11}x_{1n} & \ldots & e_{1Q} & e_{1Q}x_{11} & \ldots & e_{1Q}x_{1n} \\ e_{21} & e_{21}x_{21} & \ldots & e_{21}x_{2n} & \ldots & e_{2Q} & e_{2Q}x_{21} & \ldots & e_{2Q}x_{2n} \\ \vdots & \vdots & \vdots & \vdots & \vdots & \vdots & \vdots & \vdots & \vdots \\ e_{M1} & e_{M1}x_{M1} & \ldots & e_{M1}x_{Mn} & \ldots & e_{MQ} & e_{MQ}x_{M1} & \ldots & e_{MQ}x_{Mn} \end{pmatrix}}_{\mathbf{H}} \cdot \underbrace{\begin{pmatrix} p_0^{(1)} \\ p_1^{(1)} \\ \vdots \\ p_n^{(1)} \\ \vdots \\ p_0^{(Q)} \\ p_1^{(Q)} \\ \vdots \\ p_n^{(Q)} \end{pmatrix}}_{\vec{p}} =$$

$$= \underbrace{\begin{pmatrix} u_1 \\ \vdots \\ u_M \end{pmatrix}}_{\vec{u}}, \text{ where } e_{ij} = \frac{\mu_{\gamma_1^{(j_1)}}(x_{i1}) \times \ldots \times \mu_{\gamma_n^{(j_n)}}(x_{in})}{\sum_{k=1}^{Q} \mu_{\gamma_1^{(k_1)}}(x_{i1}) \times \ldots \times \mu_{\gamma_n^{(k_n)}}(x_{in})}.$$

Usually $M \gg (n+1) \cdot Q$, therefore this system of equations does not always have solutions, in which case it is recommended to search for a vector $\vec{p}$ instead that satisfies the following condition [28]:

$$\begin{aligned} \xi(\vec{p}) &= \|\mathbf{H} \cdot \vec{p} - \vec{u}\|^2 = (\mathbf{H} \cdot \vec{p} - \vec{u})^T \cdot (\mathbf{H} \cdot \vec{p} - \vec{u}) \to \\ \min \xi(\vec{p}) &= (\mathbf{H} \cdot \vec{p} - \vec{u})^T \cdot (\mathbf{H} \cdot \vec{p} - \vec{u}) = \\ &= (\mathbf{H} \cdot \vec{p})^T \cdot (\mathbf{H} \cdot \vec{p}) - (\mathbf{H} \cdot \vec{p})^T \cdot \vec{u} - \vec{u}^T \cdot (\mathbf{H} \cdot \vec{p}) + \\ &\quad + \vec{u}^T \cdot \vec{u} = \vec{p}^T \cdot \mathbf{H}^T \cdot \mathbf{H} \cdot \vec{p} - 2 \cdot \vec{u}^T \cdot \mathbf{H} \cdot \vec{p} + \|\vec{u}\|^2 \\ \frac{d\xi(\vec{p})}{d\vec{p}} &= 2 \cdot \mathbf{H}^T \cdot \mathbf{H} \cdot \vec{p} - 2 \cdot \vec{u}^T \cdot \mathbf{H} = 2 \cdot \mathbf{H}^T \cdot \mathbf{H} \cdot \vec{p} - 2 \cdot \mathbf{H}^T \cdot \vec{u} = \\ &= \vec{0} \mathbf{H}^T \cdot \mathbf{H} \cdot \vec{p} = \mathbf{H}^T \cdot \vec{u} \Rightarrow \vec{p} = \left(\mathbf{H}^T \cdot \mathbf{H}\right)^{-1} \cdot \mathbf{H}^T \cdot \vec{u} \end{aligned}$$

5. Calculate coefficients a_{ij}, b_{ij}, c_{ij} using the backpropagation algorithm in batch mode as follows:

$$\begin{gathered} a_{ij} := a_{ij} - \frac{\alpha_a}{M} \cdot \frac{\partial \sum_{k=1}^{M} E(\vec{x_k})}{\partial a_{ij}}, \\ b_{ij} := b_{ij} - \frac{\alpha_b}{M} \cdot \frac{\partial \sum_{k=1}^{M} E(\vec{x_k})}{\partial b_{ij}}, \\ c_{ij} := c_{ij} - \frac{\alpha_c}{M} \cdot \frac{\partial \sum_{k=1}^{M} E(\vec{x_k})}{\partial c_{ij}}, \\ 0 < \alpha_a \leqslant 1,\ 0 < \alpha_b \leqslant 1,\ 0 < \alpha_c \leqslant 1. \end{gathered}$$

6. Increase the counter of current iterations, i.e., $t := t + 1$.
7. Terminate the algorithm if one of the following conditions is satisfied: $t \geqslant T$ or $\sum_{k=1}^{M} E(\vec{x_k}) \leqslant \varepsilon$, where $E(\vec{x_k}) = \frac{1}{2} \cdot \left(u_k - y_k^{(K_{all})}\right)^2$ is an MSE of the neuro-fuzzy network, which consists of 5 layers and 1 neuron on the output layer ($K_{all} = 5, N_{K_{all}} = 1$), and has the output value $y_k^{(K_{all})}$ if vector $\vec{x_k}$ is passed to the distributed layer (with the desired output value u_k); otherwise jump to step 4.

In contrast to sequential mode, in the previous algorithm the weight adjustment was performed only after presenting the entire training sample to the ANFIS input accumulated during the epoch.

5.3.3 Support Vector Machines

Support vector machine (SVM) [29] is one of the most widely used approaches for solving the tasks of classification [30], regression [31], and prediction [32]. SVM has a simple geometric analogy, which is associated with the assumption that the elements of various classes may be linearly separated by subspaces. The main idea is to construct a linear hyperplane, which ensures the necessary partition of classified points. It is trivial that there may be more than one such hyperplanes (if any); therefore it is necessary to provide an optimization criterion that would make it possible to select the most suitable surface among all those that are suitable. A quite natural and reasonable criterion is that the margin between the points, nearest to this hyperplane and belonging to different classes, is maximized.

Figure 5.3 schematically demonstrates two elements of two classes, C_α and C_β. These elements can be separated by several different hyperplanes, which are described by the set of equations $\vec{w}^T \cdot \vec{z} - b = 0$, and differ from each other by the normal vector $\vec{w}$ (specifying the slope of a hyperplane) and bias parameter b (specifying the ascent or descent of a hyperplane). Denote the left-hand side of this equation as $y(\vec{z})$, i.e., $y(\vec{z}) = \vec{w}^T \cdot \vec{z} - b$. When substituting a specific value $\vec{z'}$ into this function, three different variants are possible. The first variant, $y(\vec{z'}) > 0$, corresponds to the case when vector $\vec{z'}$ is above the hyperplane. For the hyperplane H depicted in Fig. 5.3, we note that $y(\vec{z'}) = \vec{w}^T \cdot \vec{z'} - b > 0$ is equivalent to $\vec{z'} \in C_\alpha$. In the second variant, the inequality $y(\vec{z'}) < 0$ means that vector $\vec{z'}$ is below hyperplane H and therefore $\vec{z'} \in C_\beta$. Finally, the third variant, $y(\vec{z'}) = 0$, corresponds to the boundary case when vector $\vec{z'}$ is the decision of the equation, which means that it belongs to the separating hyperplane.

Figure 5.3 shows several separating surfaces. Two of them, H_α and H_β, refer to boundary separating hyperplanes, and are described by the equations $\vec{w_\alpha}^T \cdot \vec{z} - b_\alpha = 0$ and $\vec{w_\beta}^T \cdot \vec{z} - b_\beta = 0$, are parallel to the optimal hyperplane H_O, are an equal

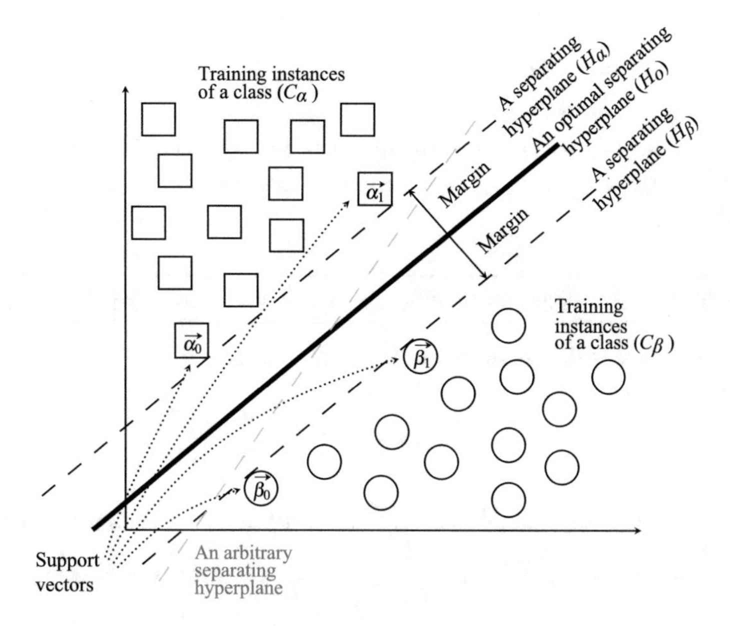

Fig. 5.3 Hyperplanes in the support vector machine

distance d from H_O, and are the nearest to the elements of classes C_α and C_β among all the other hyperplanes. The nearest elements marked on the figure as $\vec{\alpha_0}$, $\vec{\alpha_1}$, $\vec{\beta_0}$, $\vec{\beta_1}$, which are called support vectors and these vectors affect parameters $\vec{w}$ and b, which are configured during the learning process.

Such a hyperplane H_O can be described using the equation $\vec{w_O}^T \cdot \vec{z} - b_O = 0$, where $\vec{w_O}^T = (w_{O1}, \ldots, w_{On})^T$, then $\vec{w_\alpha} = \vec{w_O}$, $b_\alpha = b_O + \varepsilon$, $\vec{w_\beta} = \vec{w_O}$, $b_\beta = b_O - \varepsilon$ because of the parallelism of the planes ($\varepsilon > 0$). Without loss of generality, it can be assumed that $\varepsilon = 1$ (otherwise this can be achieved by dividing both parts of equations by ε). After simple transformations, the equations of the two separating hyperplanes, H_α and H_β, take the form $\vec{w_O}^T \cdot \vec{z} - b_O = 1$ and $\vec{w_O}^T \cdot \vec{z} - b_O = -1$, while classes C_α and C_β are presented as $C_\alpha = \left\{\vec{z} \,\middle|\, \vec{w_O}^T \cdot \vec{z} - b_O \geqslant 1\right\}$, $C_\beta = \left\{\vec{z} \,\middle|\, \vec{w_O}^T \cdot \vec{z} - b_O \leqslant -1\right\}$.

The margin value d can be calculated by dividing the distance between the boundary hyperplanes, H_α and H_β, by two, which is equal to $d_\alpha + d_\beta$. Figure 5.3 shows that the value of $d_\alpha + d_\beta$ is the length of the projection of the difference of any pair of support vectors $(\vec{\alpha_1}, \vec{\beta_1})$ from different classes onto the normal of a hyperplane $\vec{w_O}$, i.e., the scalar product of the normal $\frac{\vec{w_O}}{\|\vec{w_O}\|}$ of the optimal hyperplane H_O and the vector difference $\vec{\alpha_1} - \vec{\beta_1}$. Thus we obtain the following expression: $2 \cdot d = \frac{\vec{w_O}^T}{\|\vec{w_O}\|} \cdot \left(\vec{\alpha_1} - \vec{\beta_1}\right) = \frac{\vec{w_O}^T \cdot \left(\vec{\alpha_1} - \vec{\beta_1}\right)}{\|\vec{w_O}\|} = \frac{\vec{w_O}^T \cdot \vec{\alpha_1} - \vec{w_O}^T \cdot \vec{\beta_1}}{\|\vec{w_O}\|} = \frac{b_O + 1 - (b_O - 1)}{\|\vec{w_O}\|} = \frac{2}{\|\vec{w_O}\|}$. This yields to $d = d_\alpha = d_\beta = \frac{1}{\|\vec{w_O}\|}$. Consequently, the SVM model is described using the formula

$$Y(\vec{z}) = \text{sign}\left(\vec{w_O}^T \cdot \vec{z} - b_O\right) = \text{sign}\left(\sum_{i=1}^{n} w_{Oi} \cdot z_i - b_O\right).$$

Now we consider the training algorithm of SVM subject to the existence of the linear hyperplanes H_α and H_β, which correctly separate all training instances (see Algorithm 7).

Algorithm 7: Algorithm for training the support vector machine

1. Prepare training data in the form $\{(\vec{x_i}, u_i)\}_{i=1}^{M}$, where

$$u_i = \left[\vec{x_i} \in C_\alpha\right] - \left[\vec{x_i} \in C_\beta\right] = \begin{cases} 1, & \text{if } \vec{x_i} \in C_\alpha \\ -1, & \text{if } \vec{x_i} \in C_\beta. \end{cases}$$

2. Calculate Lagrange multipliers $\lambda_i^{(O)}$ by solving the optimization task $-\frac{1}{2} \cdot \sum_{i=1}^{M}\sum_{j=1}^{M} \lambda_i \cdot \lambda_j \cdot u_i \cdot u_j \cdot \vec{x_i}^T \cdot \vec{x_j} + \sum_{i=1}^{M} \lambda_i \longrightarrow \max_{\lambda_1,\ldots,\lambda_M}$ with the restrictions that $\sum_{i=1}^{M} \lambda_i \cdot u_i = 0$ and $\lambda_i \geqslant 0\ (i = 1, \ldots, M)$.
   To correctly classify the training object $\{\vec{x_i}\}_{i=1}^{M}$, it is necessary and sufficient that signs of the values of expressions $\left(\vec{w}^T \cdot \vec{x_i} - b\right)$ and u_i are simultaneously positive or negative. Therefore it holds that the sign of $\left(\vec{w}^T \cdot \vec{x_i} - b\right) \cdot u_i$ is greater than zero $(i = 1, \ldots, M)$. Taking into account the assumption of the possibility of the linear separation of the training instances, we note that the arguments of both multipliers on the left side is not less than 1; thus $\left(\vec{w}^T \cdot \vec{x_i} - b\right) \cdot u_i \geqslant 1$. Because the distance between the optimal separating hyperplane and the elements to classify is inversely proportional to the norm of the normal vector to this hyperplane, the maximization of this margin is equal to the optimization criteria $\Phi(\vec{w}) = \frac{1}{2} \cdot \|\vec{w}\|^2 = \frac{1}{2} \cdot \vec{w}^T \cdot \vec{w} \longrightarrow \min_{\vec{w}}$. According to the Kuhn-Tucker theorem [33], the minimization of the quadratic functional $\Phi(\vec{w})$ can be determined using the method of Lagrange multipliers as follows: $L(\vec{w}, b, \lambda_1, \ldots, \lambda_M) = \frac{1}{2} \cdot \vec{w}^T \cdot \vec{w} - \sum_{i=1}^{M} \lambda_i \cdot \left(\left(\vec{w}^T \cdot \vec{x_i} - b\right) \cdot u_i - 1\right) \longrightarrow \min_{\vec{w},b} \max_{\lambda_1,\ldots,\lambda_M}$. The necessary conditions for a saddle point $\left(\vec{w_O}, b_O, \lambda_1^{(O)}, \ldots, \lambda_M^{(O)}\right)$, i.e., the point satisfying the condition $\max_{\lambda_1,\ldots,\lambda_M} L(\vec{w_O}, b_O, \lambda_1, \ldots, \lambda_M) = L\left(\vec{w_O}, b_O, \lambda_1^{(O)}, \ldots, \lambda_M^{(O)}\right) = \min_{\vec{w},b} L\left(\vec{w}, b, \lambda_1^{(O)}, \ldots, \lambda_M^{(O)}\right)$, to exist are conditions of nullification of partial derivatives of the Lagrangian $L(\vec{w}, b, \lambda_1, \ldots, \lambda_M)$ with respect to variables $\vec{w}$, b:

$$\begin{cases} \frac{\partial L(\vec{w},b,\lambda_1,\ldots,\lambda_M)}{\partial \vec{w}} = \vec{0} \\ \frac{\partial L(\vec{w},b,\lambda_1,\ldots,\lambda_M)}{\partial b} = 0 \end{cases} \Rightarrow \begin{cases} \vec{w} - \sum\limits_{i=1}^{M} \lambda_i \cdot u_i \cdot \vec{x}_i = \vec{0} \\ \sum\limits_{i=1}^{M} \lambda_i \cdot u_i = 0. \end{cases}$$

The Lagrangian takes the following form:

$L(\vec{w}, b, \lambda_1, \ldots, \lambda_M) = \frac{1}{2} \cdot \vec{w}^T \cdot \vec{w} - \sum\limits_{i=1}^{M} \lambda_i \cdot u_i \cdot \vec{w}^T \cdot \vec{x}_i + b \cdot \sum\limits_{i=1}^{M} \lambda_i \cdot u_i +$
$\sum\limits_{i=1}^{M} \lambda_i = \frac{1}{2} \cdot \sum\limits_{i=1}^{M} \sum\limits_{j=1}^{M} \lambda_i \cdot \lambda_j \cdot u_i \cdot u_j \cdot \vec{x}_i^{\,T} \cdot \vec{x}_j - \sum\limits_{i=1}^{M} \sum\limits_{j=1}^{M} \lambda_i \cdot \lambda_j \cdot u_i \cdot u_j \cdot \vec{x}_i^{\,T} \cdot \vec{x}_j$
$+ \sum\limits_{i=1}^{M} \lambda_i = -\frac{1}{2} \cdot \sum\limits_{i=1}^{M} \sum\limits_{j=1}^{M} \lambda_i \cdot \lambda_j \cdot u_i \cdot u_j \cdot \vec{x}_i^{\,T} \cdot \vec{x}_j + \sum\limits_{i=1}^{M} \lambda_i$. The required values $\lambda_i^{(O)}$ $(i = 1, \ldots, M)$ are calculated by the Lagrangian maximization $L(\lambda_1, \ldots, \lambda_M) \equiv L(\vec{w}, b, \lambda_1, \ldots, \lambda_M)$ with the restrictions $\sum_{i=1}^{M} \lambda_i \cdot u_i = 0$ and $\lambda_i \geqslant 0$ $(i = 1, \ldots, M)$. For this, one can apply sequential minimal optimization [34].

3. Calculate the normal vector and free coefficient in the optimal separating hyperplane equation. Vector $\vec{w_O}$ is represented as $\vec{w_O} = \sum\limits_{i=1}^{M} \lambda_i^{(O)} \cdot u_i \cdot \vec{x}_i$. At the saddle point $(\vec{w_O}, b_O)$, according to the complementary slackness condition, the following identity holds: $\lambda_i^{(O)} \cdot \left(\left(\vec{w_O}^T \cdot \vec{x}_i - b_O \right) \cdot u_i - 1 \right) = 0$, where $i = 1, \ldots, M$. The $\vec{x}_{i_j}$ vectors, for which non-zero $\lambda_{i_j}^{(O)}$ exists $(i_j \in \{i_1, \ldots, i_{\widetilde{M}}\} \subseteq \{1, \ldots, M\})$, are called support vectors, which satisfy the equation of one of the boundary separating hyperplanes (H_α or H_β). Based on this, we can calculate the coefficient $b_O = \vec{w_O}^T \cdot \vec{x}_{i_j} - u_{i_j}$, where $\vec{x}_{i_j}$ is any of support vectors. Thereby the normal vector $\vec{w_O}$ is represented as a linear combination of support vectors $\vec{x}_{i_j}$ $(j = 1, \ldots, \widetilde{M})$.

4. Clarify the SVM model as follows:

$$Y(\vec{z}) = \text{sign}\left(\sum_{j=1}^{\widetilde{M}} w_{i_j} \cdot \vec{x}_{i_j}^{\,T} \cdot \vec{z} - b_O \right),$$

where $w_{i_j} = \lambda_{i_j}^{(O)} \cdot u_{i_j}$, and the summation operator is applied to the index subset $\{i_1, \ldots, i_{\widetilde{M}}\}$ of the training sample, which corresponds to support vectors.

If the objects from different classes cannot be separated linearly, two methods may be used, both of which are aimed at decreasing the empirical risk on the elements of the training set. The first approach is to apply special transformations, namely, kernels φ for transition to a new $\widehat{n}$-dimensional space [35]. It is supposed that in a new space, which is a result of the mapping $\vec{\varphi}(\vec{z}) = (\varphi_1(\vec{z}), \ldots, \varphi_{\tilde{n}}(\vec{z}))^T$, there is a hyperplane satisfying the previously specified criteria. The second approach is based on introducing a penalty function, which allows us to ignore some of the incorrectly classified objects based either on the total quantity or on the total distance from the separating hyperplane. The first case means searching for such a separating hyperplane, which provides a minimum value of the characteristic function $\sum_{i=1}^{M} \left[Y(\vec{x}_i) \neq u_i\right]$. In the second case, the objective function is $\sum_{i=1}^{M} \left[Y(\vec{x}_i) \neq u_i\right] \cdot \mathrm{dist}(\vec{x}_i, H_O)$, where $\mathrm{dist}(\cdot, \cdot)$ is the distance function between the indicated (vector, hyperplane) argument pairs within the given metrics.

5.4 Training the Binary Classifier for Detecting Network Attacks

The following sections discuss how to calculate network parameters and optimize the weights of the binary classifier, and introduces an algorithm for detecting network attacks.

5.4.1 Calculating and Preprocessing Network Parameters

Our proposed approach for detecting network attacks using a binary classifier consists of three steps. The first step is to calculate the network parameters and preprocess them. The second step is genetic optimization. The third step is detecting network attacks using a binary classifier.

We have developed a network analyzer, which allows us to construct 106 parameters to describe the network connections between hosts. These parameters include connection duration, network service, the intensity of sending special packets, the number of active connections between concrete IP address pairs (one of the criteria of DoS attacks), the binary characteristic of changing the TCP[5] window scale after establishing an actual session, the current state of the TCP connection, different attributes of the presence of scanning packet at the levels TCP, UDP,[6] ICMP,[7] IP,[8] etc.

To calculate the statistical parameters of network attacks, we have used the adapted sliding window method. This method basically splits a given time interval $\Delta_0^{(L)} = [0, L]$ with length L, during which a continuous observation of a number of parameters is carried out, into several smaller intervals $\Delta_0^{(L')}, \Delta_\delta^{(L')}, \ldots, \Delta_{\delta\cdot(k-1)}^{(L')}$ with identical length $0 < L' \leqslant L$, the beginning of each has an offset $0 < \delta \leqslant L'$ relative

[5]Transmission Control Protocol.

[6]User Datagram Protocol.

[7]Internet Control Message Protocol.

[8]Internet Protocol.

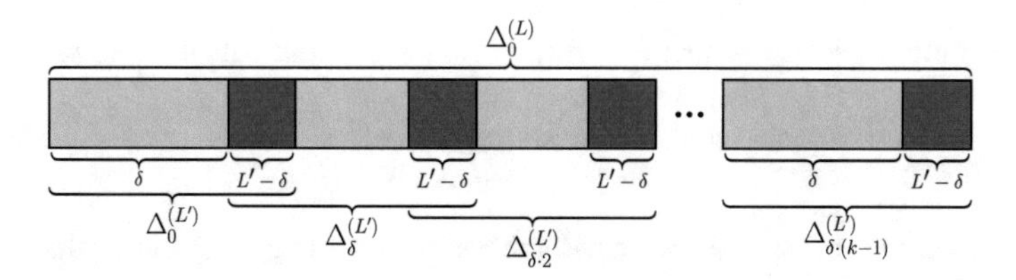

Fig. 5.4 Sliding window method

to the beginning of the previous interval (see Fig. 5.4). If $\bigcup_{i=0}^{k-1} \Delta_{\delta\cdot i}^{L'} \subseteq \Delta_0^{(L)}$ and $\bigcup_{i=0}^{k} \Delta_{\delta\cdot i}^{L'} \supsetneqq \Delta_0^{(L)}$, then $k = 1 + \lfloor \frac{L-L'}{\delta} \rfloor$. During time intervals $\Delta_0^{(L')}, \ldots, \Delta_{\delta\cdot(k-1)}^{(L')}$, the snapshots of the parameter values $\omega_0, \ldots \omega_{k-1}$ are made and their average value (intensity) $\bar{\omega}$ within the time window with length L' is calculated by the formula $\bar{\omega} = \frac{1}{k} \cdot \sum\limits_{i=0}^{k-1} \omega_i$. In the experiments, we have used an interval whose value for parameter L was five seconds. The length of the smoothing interval L' was chosen to be 1 s. The offset δ was set to half a second. Supposedly, such approach eliminates sparse and occasional network bursts, and decreases the false positive rate.

For preprocessing, we have used principal component analysis (PCA) [36], which is described as a sequence of the steps described in Algorithm 8.

Algorithm 8: Principal component analysis

1. Calculate the mathematical expectation of the random vector, which in our case is represented by the set of training elements
$\left\{\vec{x}_i = \left\{x_{ij}\right\}_{j=1}^{n}\right\}_{i=1}^{M}$: $\bar{\vec{x}} = (\bar{x}_1, \ldots, \bar{x}_n)^T = E\left[\{\vec{x}_i\}_{i=1}^{M}\right] = \frac{1}{M} \cdot \sum\limits_{i=1}^{M} \vec{x}_i = \left(\frac{1}{M} \cdot \sum\limits_{i=1}^{M} x_{i1}, \ldots, \frac{1}{M} \cdot \sum\limits_{i=1}^{M} x_{in}\right)^T$.

2. Generate elements of an unbiased theoretical covariance matrix $\Sigma =$ $\left(\sigma_{ij}\right)_{\substack{i=1,\ldots,n \\ j=1,\ldots,n}}$:
$$\sigma_{ij} = \frac{1}{M-1} \cdot \sum_{k=1}^{M} (x_{ki} - \bar{x}_i) \cdot \left(x_{kj} - \bar{x}_j\right).$$

3. Calculate eigenvalues $\{\lambda_i\}_{i=1}^{n}$ and eigenvectors $\{\vec{\nu}_i\}_{i=1}^{n}$ of the matrix Σ as the roots of equations (for this purpose we have used the Jacobi rotation in relation to matrix Σ):
$$\begin{cases} \det(\Sigma - \lambda \cdot \mathbf{I}) = 0 \\ (\Sigma - \lambda \cdot \mathbf{I}) \cdot \vec{\nu} = \vec{0}. \end{cases}$$

4. Sort eigenvalues $\{\lambda_i\}_{i=1}^{n}$ in decreasing order and corresponding eigenvectors $\{\vec{\nu}_i\}_{i=1}^{n}$:
$$\lambda_1 \geqslant \lambda_2 \geqslant \ldots \lambda_n \geqslant 0.$$

5. Select the required number $\widehat{n} \leqslant n$ of principal components as follows:

$$\widehat{n} = \min \{j | \varsigma(j) \geqslant \varepsilon\}_{j=1}^{n},$$

where $\varsigma(j) = \frac{\sum_{i=1}^{j} \lambda_i}{\sum_{i=1}^{n} \lambda_i}$ is an informativeness measure, $0 \leqslant \varepsilon \leqslant 1$ is a value selected expertly;
6. Center the input feature vector $\vec{z}$ as $\vec{z}_c = \vec{z} - \bar{\vec{x}}$.
7. Project the centered feature vector $\vec{z}_c$ into the new coordinate system, described by the orthonormalized vectors $\{\vec{\nu}_i\}_{i=1}^{\widehat{n}}$:

$$\vec{y} = (y_1, \ldots, y_{\widehat{n}})^T = (\vec{\nu}_1, \ldots, \vec{\nu}_{\widehat{n}})^T \cdot \vec{z}_c,$$

where $y_i = \vec{\nu}_i^{\,T} \cdot \vec{z}_c$ is called ith principal component of vector $\vec{z}$.

Figure 5.5 demonstrates the dependence of the informativeness measure ς on the number of principal components.

In our experiments, we have compressed the dimensionality of the input vectors to 33, which corresponds to $\varepsilon = 0.99$.

5.4.2 *Genetic Optimization of the Weights of the Binary Classifier*

To accelerate the training process, we use genetic operators such as crossover G_1, mutation G_2, and permutation G_3. After several training epochs, the inner weights of the adaptive classifiers are tuned. To this end, we create two copies of the original classifier in which some weight vectors have been modified. The rule for genetic optimization of weights within the classifier is represented as follows:

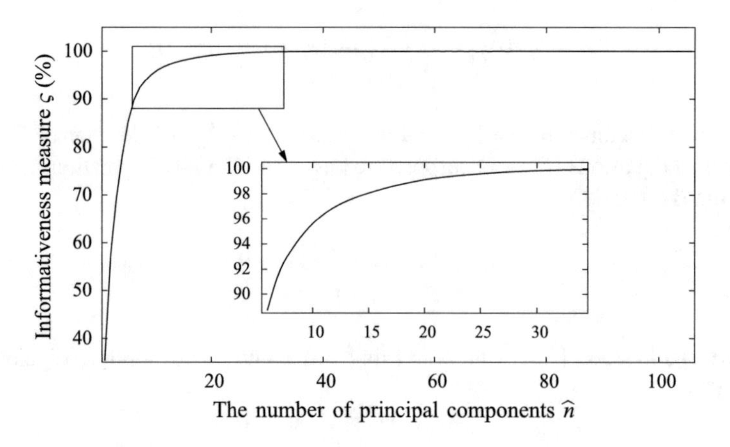

Fig. 5.5 Informativeness measure in PCA

Algorithm 9: Genetic optimization of weights of the binary classifier

1. Clone the binary classifier B in two copies, B' and B''.
2. Randomly select one of three genetic operators, e.g., $G \in \{G_1, G_2, G_3\}$.
3. Randomly select two neurons (τ' and τ'') located in the same layer.
4. Apply the genetic operator G (one of the steps 4a, 4b, or 4c) to weight vectors $\overrightarrow{w'} = \left(w'_1, \ldots, w'_k, w'_{k+1}, \ldots, w'_n\right)^T$ and $\overrightarrow{w''} = \left(w''_1, \ldots, w''_k, w''_{k+1}, \ldots, w''_n\right)^T$ to neurons τ' and τ''.

   a. If operator G is a crossover, split vectors $\overrightarrow{w'}$ and $\overrightarrow{w''}$ into two portions adjustable via parameter k ($1 \leqslant k < n$ is a randomly selected number) and swap these portions as follows: $\overrightarrow{w'} = \left(w''_1, \ldots, w''_k, w'_{k+1}, \ldots, w'_n\right)^T$ and $\overrightarrow{w''} = \left(w'_1, \ldots, w'_k, w''_{k+1}, \ldots, w''_n\right)^T$ (see Fig. 5.6).

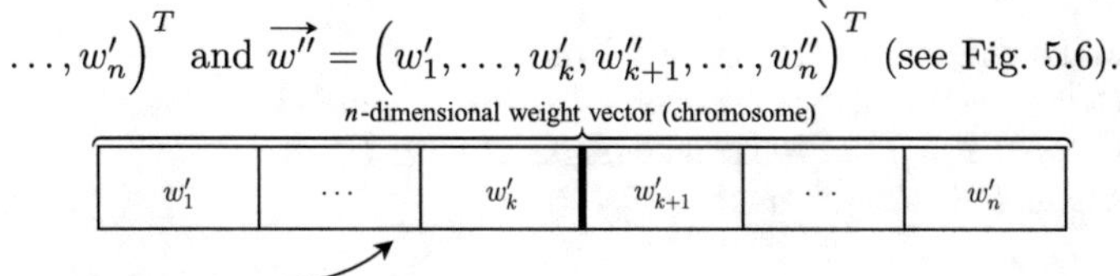

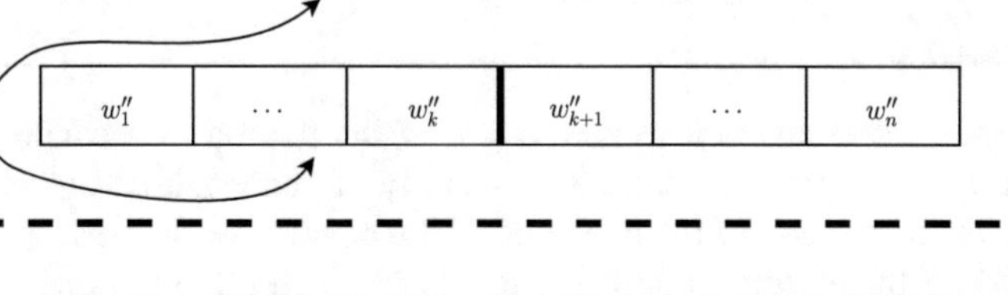

Fig. 5.6: Crossover G_1

   b. If operator G is a mutation, swap a pair of randomly selected and mismatched bits $b'_{p'}$ and $b'_{q'}$ within gene $w'_{k'}$; repeat the same for gene $w''_{k''}$ (see Fig. 5.7).

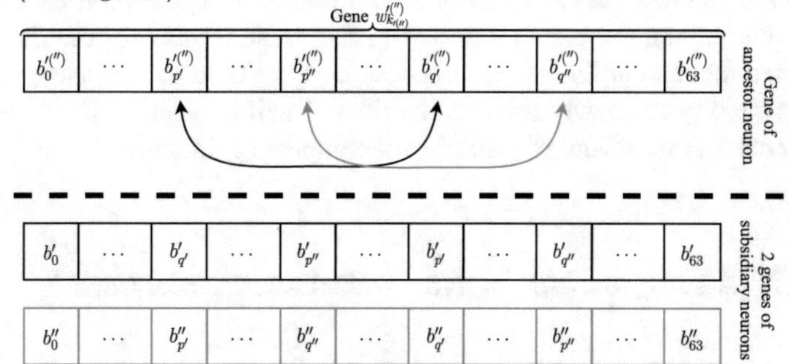

Fig. 5.7: Mutation G_2

c. If operator G is a permutation, inverse a randomly selected bit $b'_{p'}$ within gene $w'_{k'}$; repeat the same for gene $w''_{k''}$ (see Fig. 5.8).

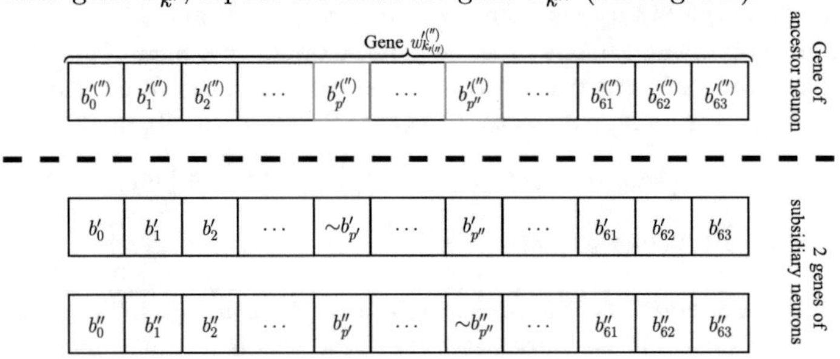

Fig. 5.8: Permutation G_3

5. Make changes to classifiers B' and B''.
6. Calculate the fitness functions F_B, $F_{B'}$, $F_{B''}$ of classifiers B, B', and B''.
7. If $F_B < F_{B^*}$, then replace classifier B with B^*, where $B^* \in \operatorname{Arg} \max_{\bar{B} \in \{B', B''\}} F_{\bar{B}}$.

Here we represent each weight vector of the neuron as a single chromosome, which consists of several genes. The quantity of genes within this chromosome is equal to the dimension of the layer before the position of the corresponding neuron (encoded by this chromosome). From step 4a, it can be seen that the bit content of genes persists inside descendants while applying the crossover: the boundary specifying the separation of the chromosome passes between adjacent genes without violating their integrity. At the same time, the operators of mutation and permutation (steps 4b and 4c) are applied only to the part of the selected gene corresponding to the mantissa of a 64-bit real number. Such restriction is due in order to avoid an explosive growth of the gene content. Step 6 is performed for each classifier in a separate thread: after making the genetic corrections on the child classifiers, it is necessary to calculate the level of their suitability. First, we perform parallel training of the child classifiers along with their ancestor during several epochs. From these three classifiers, the one that possesses the greatest value of the fitness function is selected. Instead of the fitness function, here we used the inverse of the MSE, E^{-1}. Subsequently the newly formed classifier B will be used as the parent to generate other classifiers, B' and B'', which will compete again with B.

5.4.3 An Algorithm for Network Attack Detection

The algorithm for classifying network attacks using the binary classifier $\widetilde{Y}$ can be represented as shown in Algorithm 10.

Algorithm 10: Network attack detection algorithm

1. Select a binary classifier (detector) $\widetilde{Y}$ and specify its parameters.
2. Choose the class sets $\widetilde{C_{\widetilde{ij}}} = \widetilde{C_{\widetilde{i}}} \bigvee \widetilde{C_{\widetilde{j}}} \subseteq C$ for training the detector $\widetilde{Y}$.
3. Prepare the training data $\Upsilon_{\mathcal{X}^{(LS)}_{\widetilde{C}_{\widetilde{ij}}}} = \left\{\left(\vec{x_{\widetilde{k}}}, \bar{c_{\widetilde{k}}}\right)\right\}_{\widetilde{k}=1}^{\widetilde{M}}$, where

$$\widetilde{M} = \#\mathcal{X}^{(LS)}_{\widetilde{C}_{\widetilde{ij}}} \text{ is the cardinality of the training set for detector } \widetilde{Y},$$

$$\bar{c_{\widetilde{k}}} = \begin{cases} \widetilde{i}, & \text{if } u_{\widetilde{k}} \in \{-1, 0\} \\ \widetilde{j}, & \text{if } u_{\widetilde{k}} = 1 \end{cases} \quad \text{are class labels, and}$$

$$u_{\widetilde{k}} = \begin{cases} \bar{y_{\widetilde{i}}} = \begin{bmatrix} -1 \\ 0 \end{bmatrix}, & \text{if } \exists \widetilde{C} \in \widetilde{C_{\widetilde{i}}}\, \vec{x_{\widetilde{k}}} \in \widetilde{C} \\ \bar{y_{\widetilde{j}}} = 1, & \text{if } \exists \widetilde{C} \in \widetilde{C_{\widetilde{j}}}\, \vec{x_{\widetilde{k}}} \in \widetilde{C} \end{cases} \quad \text{is the desired output.}$$

The parameter value $\bar{y_{\widetilde{i}}}$ is chosen to be -1 or 0, depending on a type of the classifier $\widetilde{Y}$ and the activation/membership function on its output layer.
4. Train the binary classifier $\widetilde{Y}$ using data $\Upsilon_{\mathcal{X}^{(LS)}_{\widetilde{C}_{\widetilde{ij}}}}$.
5. Classify the input object $\vec{z}$ represented as a feature vector: if $\widetilde{Y}(\vec{z}) < \bar{h}_{\widetilde{ij}}$, then object $\vec{z}$ belongs to one of the classes $\widetilde{C_{\widetilde{i}}}$; if $\widetilde{Y}(\vec{z}) \geqslant \bar{h}_{\widetilde{ij}}$, then object $\vec{z}$ belongs to one of the classes $\widetilde{C_{\widetilde{j}}}$. Here $\bar{h}_{\widetilde{ij}} = \frac{\bar{y_{\widetilde{i}}} + \bar{y_{\widetilde{j}}}}{2}$ can be interpreted as an activation threshold of detector $\widetilde{Y}$. More precisely, by denoting C_0 as a class of normal connections, if we have $\widetilde{C_{\widetilde{i}}} = \{C_0\} \wedge \widetilde{Y}(\vec{z}) < \bar{h}_{\widetilde{ij}}$, then the detector $\widetilde{Y}$ recognizes object $\vec{z}$ as normal.

5.5 Schemes for Combining the Binary Classifiers

This section discusses how to combine detectors using low-level schemes and aggregate compositions.

5.5.1 *Low-Level Schemes for Combining Detectors*

In this section we consider several schemes that combine the binary classifiers and are designed for associating an object with one of the $(m + 1)$ class labels.

In the one-against-all approach, each detector $F^{(i)}_{jk} : \mathbb{R}^n \to \{0, 1\}$ $(k = 1, \ldots, m)$ is trained on data $\{(\vec{x_l}, [\bar{c_l} = k])\}_{l=1}^{M}$, and functioning the group of detectors $F^{(i)}_j$ is described using the exclusionary principle:

$$F_j^{(i)}(\vec{z}) = \begin{cases} \{0\} & \text{if } \forall k \in \{1, \ldots, m\} \ F_{jk}^{(i)}(\vec{z}) = 0 \\ \left\{ k \left| F_{jk}^{(i)}(\vec{z}) = 1 \right. \right\}_{k=1}^{m} & \text{otherwise.} \end{cases}$$

In the one-against-one approach, each $\binom{m+1}{2} = \frac{(m+1)\cdot m}{2}$ detectors $F_{jk_0k_1}^{(i)}$ is trained on the set of objects that belong to only two classes, k_0 and k_1, namely, using the set $\{(\vec{x}_l, 0 \,|\bar{c}_l = k_0)\}_{l=1}^{M} \bigcup \{(\vec{x}_l, 1 \,|\bar{c}_l = k_1)\}_{l=1}^{M}$, where $0 \leqslant k_0 < k_1 \leqslant m$, and the group of detectors can be set using max-wins voting as follows:

$$F_j^{(i)} = \left\{ \arg \max_{\bar{c} \in \{0,\ldots,m\}} \sum_{k=\bar{c}+1}^{m} \left[F_{j\bar{c}k}^{(i)}(\vec{z}) = 0 \right] + \sum_{k=0}^{\bar{c}-1} \left[F_{jk\bar{c}}^{(i)}(\vec{z}) = 1 \right] \right\}.$$

One of the variations of the previous approaches for combining the detectors is the classification binary tree. Formally speaking, this structure is recursively defined as follows:

$$CBT_\mu = \begin{cases} \langle F_{jL_\mu R_\mu}^{(i)}, CBT_{L_\mu}, CBT_{R_\mu} \rangle, & \text{if } \#\mu \geqslant 2 \\ \mu, & \text{if } \#\mu = 1. \end{cases}$$

Here $\mu = \{0, \ldots, m\}$ is the original set of class labels, $L_\mu \subsetneq \mu$ is a randomly generated or user-defined subset of μ ($\#L_\mu < \#\mu$), $R_\mu = \mu \setminus L_\mu$, CBT_{L_μ} is a left classification binary subtree, CBT_{R_μ} is a right classification binary subtree, and $F_{jL_\mu R_\mu}^{(i)}$ is a node detector trained on elements $\left\{(\vec{x}_l, 0) \left|\bar{c}_l \in L_\mu \right.\right\}_{l=1}^{M} \bigcup \left\{(\vec{x}_l, 1) \left|\bar{c}_l \in R_\mu \right.\right\}_{l=1}^{M}$, i.e., the detector output is configured to be 0 if the input object $\vec{x}_l$ has label $\bar{c}_l \in L_\mu$, and 1 if the input object X_l has label $\bar{c}_l \in R_\mu$. Therefore the group of detectors nested in each other is given via the recursive function $\phi_j^{(i)}$, which specifies sequential fragmentation of the set μ as follows:

$$F_j^{(i)} = \phi_j^{(i)}(\mu, \vec{z}),$$

$$\phi_j^{(i)}(\mu, \vec{z}) = \begin{cases} \mu, & \text{if } \#\mu = 1 \\ \phi_j^{(i)}\left(L_\mu, \vec{z}\right), & \text{if } \#\mu \geqslant 2 \wedge F_{jL_\mu R_\mu}^{(i)}(\vec{z}) = 0 \\ \phi_j^{(i)}\left(R_\mu, \vec{z}\right), & \text{if } \#\mu \geqslant 2 \wedge F_{jL_\mu R_\mu}^{(i)}(\vec{z}) = 1. \end{cases}$$

Using function $\phi_j^{(i)}$, an unambiguous search can be performed on the class label of the input object. This is possible because if a disjoint partition of the set of class labels takes place at each step during the descent down the classification tree, then only one possible class label remains after reaching the terminal detector. Therefore conflict cases of classification through the classification tree are not possible within group $F_j^{(i)}$, although they may occur for the other two approaches considered above.

Another approach is a directed acyclic graph that combines $\binom{m+1}{2} = \frac{(m+1)\cdot m}{2}$ detectors into a linked dynamic structure, which can be given by the following formula:

$$DAG_{\mu} = \begin{cases} \langle F^{(i)}_{j\mu k_0 k_1}, DAG_{\mu \setminus \{k_0\}}, DAG_{\mu \setminus \{k_1\}} \rangle, & \text{if } \#\mu \geqslant 2, \text{ where } k_0 \in \mu, \\ & k_1 \in \mu \\ \mu, & \text{if } \#\mu = 1. \end{cases}$$

As with the one-against-one approach, each node detector $F^{(i)}_{j\mu k_0 k_1}$ is trained based on the elements $\{(\vec{x}_l, 0 \,|\bar{c}_l = k_0)\}_{l=1}^{M} \bigcup \{(\vec{x}_l, 1 \,|\bar{c}_l = k_1)\}_{l=1}^{M}$ $(k_0 < k_1)$. Bypassing the graph is performed using a recursive function $\xi^{(i)}_j$, which specifies an element-by-element detachment of set μ:

$$F^{(i)}_j = \xi^{(i)}_j (\mu, \vec{z}),$$

$$\xi^{(i)}_j (\mu, \vec{z}) = \begin{cases} \mu, & \text{if } \#\mu = 1 \\ \xi^{(i)}_j (\mu \setminus \{k_1\}, \vec{z}), & \text{if } \#\mu \geqslant 2 \wedge F^{(i)}_{j\mu k_0 k_1} (\vec{z}) = 0 \\ \xi^{(i)}_j (\mu \setminus \{k_0\}, \vec{z}), & \text{if } \#\mu \geqslant 2 \wedge F^{(i)}_{j\mu k_0 k_1} (\vec{z}) = 1. \end{cases}$$

If detector $F^{(i)}_{j\mu k_0 k_1}$ votes for the k_0th class of the object $\vec{z}$, i.e., $F^{(i)}_{j\mu k_0 k_1} (\vec{z}) = 0$, then label k_1 is removed from set μ as obviously false, otherwise label k_0 is excluded. The process is repeated until set μ degenerates into a single-element set.

Table 5.1 shows the characteristics of the considered schemes for combining detectors in a multi-class model, which attempts to associate an input object with one or more $(m + 1)$ class labels.

Only one, namely the classification binary tree, has a variable number of detectors, which can be used for classifying objects. The minimum value is reached when detector $F^{(i)}_{jL_\mu R_\mu}$ is activated, which is located in the tree root and trained to recognize only one object among all the rest and $F^{(i)}_{jL_\mu R_\mu} (\vec{z}) = 0$ ($F^{(i)}_{jL_\mu R_\mu} (\vec{z}) = 1$), i.e., when $\#L_\mu = 1$ ($\#R_\mu = 1$). The maximum value is achieved when the tree is represented by a sequential list and the most remote detector is activated in it. In the case of a balanced tree, this indicator can be equal to $\lfloor \log_2 (m + 1) \rfloor$ or $\lceil \log_2 (m + 1) \rceil$.

Table 5.1 Characteristics of the schemes for combining detectors

Scheme for combining detectors	Number of detectors to be trained	The minimum number of detectors involved in the classification of objects	The minimum number of detectors involved in the classification of objects
One-against-all	m	m	m
One-against-one	$\frac{(m+1)\cdot m}{2}$	$\frac{(m+1)\cdot m}{2}$	$\frac{(m+1)\cdot m}{2}$
Classification binary tree	m	1	m
Directed acyclic graph	$\frac{(m+1)\cdot m}{2}$	m	m

5.5.2 Aggregating Compositions

In this section, we present four aggregating compositions. Under the notation $\left\{F^{(j)}\right\}_{j=1}^{P}$ we mean basic classifiers.

1. Majority voting can be represented as follows:

$$G(\vec{z}) = \left\{ k \left| \sum_{j=1}^{P} \left[k \in F^{(j)}(\vec{z})\right] > \frac{1}{2} \cdot \sum_{i=0}^{m} \sum_{j=1}^{P} \left[i \in F^{(j)}(\vec{z})\right] \right. \right\}_{k=0}^{m}.$$

As a result of applying function G to argument $\vec{z}$, the set of labels is formed for those classes for which more than half of the votes was given.

2. An extension of the previous function is weighted voting, which is described as $G(\vec{z}) =$

$$= \left\{ k \left| \sum_{j=1}^{P} \omega_j \cdot \left[k \in F^{(j)}(\vec{z})\right] = \max_{i \in \{0,\dots,m\}} \sum_{j=1}^{P} \omega_j \cdot \left[i \in F^{(j)}(\vec{z})\right] \right. \right\}_{k=0}^{m}.$$

The coefficients ω_j are chosen in such a way that $\sum_{j=1}^{P} \omega_j = 1$ is satisfied. These can be calculated, for example, with the following formula:

$$\omega_j = \frac{\#\left\{\vec{x}_k \left| \bar{c}_k \in F^{(j)}(\vec{x}_k)\right.\right\}_{k=1}^{M}}{\sum_{i=1}^{P} \#\left\{\vec{x}_k \left| \bar{c}_k \in F^{(i)}(\vec{x}_k)\right.\right\}_{k=1}^{M}}.$$

In this formula, each coefficient ω_j $(j = 0, \dots, P)$ is a proportion of training vectors correctly (and potentially with conflicts) recognized by classifier $F^{(j)}$ among all other training vectors, which are correctly recognized by classifiers $F^{(1)}, \dots, F^{(P)}$. When substituting the values $\omega_j = \frac{1}{P}$ into the weighted voting formula, we obtain its specific analogue: simple voting (max-wins voting).

3. Stacking is represented as a composition of some function F with basic classifiers $F^{(1)}, \dots, F^{(P)}$ and identical function ID and specified using the following formula[9]:

$$G(\vec{z}) = F\left(F^{(1)}(\vec{z}), \dots, F^{(P)}(\vec{z}), \vec{z}\right).$$

In this formula, function F aggregates input vector $\vec{z}$, and output results of functions $F^{(1)}, \dots, F^{(P)}$.

4. The Fix and Hodges method is based on the idea of forming a competence area for each classifier [37]. Within this area, the upper-level classifier G fully trusts

[9]For the sake of simplicity, some features related to the training of classifiers F and $F^{(1)}, \dots, F^{(P)}$ on various subsamples are omitted here.

the corresponding basic classifier. The choice of solution (resulting in the class label) by classifier G depends on how correctly a basic classifier recognizes $\widetilde{M} \leqslant M$ trainings instances, which are located in the nearest vicinity of vector $\vec{z}$. Consequently, function G is specified as follows:

$$G(\vec{z}) = \bigcup_{l=1}^{P} \left(F^{(l)}(\vec{z}) \;\middle|\; \underbrace{\sum_{j=1}^{\widetilde{M}} \left[F^{(l)}(\vec{x}_{i_j}) = \left\{ \bar{c}_{i_j} \right\} \right]}_{D_{\widetilde{M}}(l)} = \max_{k \in \{1,\ldots,P\}} D_{\widetilde{M}}(k) \right),$$

$$\left\{ i_1, \ldots, i_{\widetilde{M}} \right\} = \arg \min_{\substack{\widetilde{I} \subseteq \{1,\ldots,M\} \\ (\#\widetilde{I} = \widetilde{M})}} \sum_{i \in \widetilde{I}} \rho\left(\vec{x}_i, \vec{z}\right).$$

In this formula, the labels of those classes are combined for which the basic classifiers have voted with the greatest number of correctly recognized training objects the distance from which to vector $\vec{z}$ is minimal in accordance with metric ρ. To this end, the index subset $\left\{ i_1, \ldots, i_{\widetilde{M}} \right\}$ of the training sample $\{(\vec{x}_i, \bar{c}_i)\}_{i=1}^{M}$ is calculated and vectors with numbers of such a subset satisfy the specified requirement. The modifications can be represented as the following formulae:

$$G(\vec{z}) = \bigcup_{l=1}^{P} \left(F^{(l)}(\vec{z}) \;\middle|\; \underbrace{\sum_{j=1}^{\widetilde{M}} \frac{\left[F^{(l)}(\vec{x}_{i_j}) = \left\{ \bar{c}_{i_j} \right\} \right]}{\rho\left(\vec{x}_{i_j}, \vec{z}\right)}}_{D^{*}_{\widetilde{M}}(l)} = \max_{k \in \{1,\ldots,P\}} D^{*}_{\widetilde{M}}(k) \right),$$

$$G(\vec{z}) = \bigcup_{l=1}^{P} \left(F^{(l)}(\vec{z}) \;\middle|\; \underbrace{\sum_{\substack{j \in \{1,\ldots,M\} \\ \rho(\vec{x}_j, \vec{z}) \leqslant \widetilde{r}}} \left[F^{(l)}(\vec{x}_j) = \left\{ \bar{c}_j \right\} \right]}_{D^{**}_{\widetilde{r}}(l)} = \max_{k \in \{1,\ldots,P\}} D^{**}_{\widetilde{r}}(k) \right).$$

In the first formula, the weight encouragement of classifiers is performed, which is directly proportional to the proximity of the correctly classified training object and test object $\vec{z}$. In the second formula, the summation is carried out only on the instances located in the vicinity with specified radius $\widetilde{r}$ from test object $\vec{z}$.

5.5.3 Common Approach for Combining Detectors

This section presents a novel approach for combining the detectors, which allows the construction of multilevel schemes. Figure 5.9 presents an example of combining the detectors (this was used during the experiments described in the next section).

The detectors are designated as $F_{ij}^{(k)}$ and combined into a group $F_i^{(k)}$ using one of the low-level schemes considered earlier in Sect. 5.5.1. Detectors within different groups are trained using various subsamples from an initial training set. As shown in the figure, the number of detectors within each group is equal to m, which corresponds to the case of the one-against-all scheme, and the number of groups for each kth basic classifier is denoted as q_k. The combination of groups $\left\{F_i^{(k)}\right\}_{i=1}^{q_k}$ into the basic classifier $F^{(k)}$ is performed using the following hybrid rule represented as a mixture of majority voting and max-wins voting:

$$F^{(k)}(\vec{z}) = \left\{ \bar{c} \,\middle|\, \underbrace{\sum_{j=1}^{q_k} \left[\bar{c} \in F_j^{(k)}(\vec{z}) \right]}_{\Xi_k(\bar{c})} > \frac{1}{2} \cdot q_k \wedge \Xi_k(\bar{c}) = \max_{\bar{c}' \in \{0,\ldots,m\}} \Xi_k\left(\bar{c}'\right) \right\}_{\bar{c}=0}^{m}.$$

It should be noted that due to the technique described in [4], it becomes possible to construct the multilevel schemes without a strict binding to the aggre-

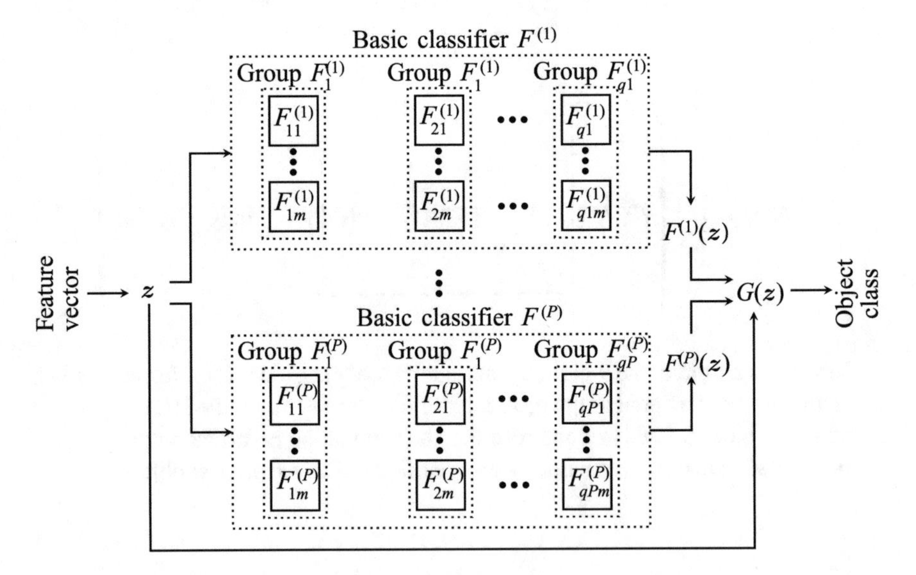

Fig. 5.9 Combining the detectors via the scheme one-against-all

gating composition. In such schemes the upper-level classifier combines the basic classifiers, which is independent of the low-level scheme for combining the detectors (i.e., binary classifiers). Moreover, the implemented interpreter and algorithm of cascade learning of classifiers allow to perform the construction of the universal structure for storage and presentation of classifiers, namely, classifier tree, and tune the node classifiers.

5.6 Experiments

In the following sections, two experiments are described, along with the dataset that was used with them.

5.6.1 The Dataset

This section describes an experimental dataset and a method of calculating performance indicators based on a 5-fold cross-validation.

For experiments, we have used the DARPA 1998 dataset.[10] This dataset is widely used for investigating methods for network attack detection, however, it is subject to severe criticism due to its synthetic nature [38]. In particular, paper [39] outlines that classification between normal and anomalous traffic could be carried out using solely IP-header fields (e.g., TTL). We have extracted more than 100 parameters, but among them there were no such fields, because their values did not reflect the presence or absence of anomaly (for example packets within a TCP-session may be transmitted on different routes in real-world networks). The parameters considered in this research are mostly statistical, and the DARPA 1998 dataset is used for experiments because of convenience: the CSV-files containing the class labels are available along with pcap-files. Moreover, the outdated attack types were excluded (e.g., buffer overflow in mail service or attack teardrop realizing the old vulnerability inside the TCP/IP stack during the defragmentation of packets), we have dealt with the anomalies that are characterized by some deviations in statistical parameters. This way, seven classes were selected: six anomalous classes, namely neptune, smurf (DoS), ipsweep, nmap, portsweep, and satan (Probe); and a normal class.

While forming the dataset, we have used the binary network traces "Training Data Set 1998," which were collected on the Wednesday of the first week, on the Monday of the second week, on the Tuesday of the second week, on the Monday of the third week, on the Wednesday of the third week, on the Friday of the third week, on the Tuesday of the fourth week, and on the Wednesday of the fourth week of the experiment.

---

[10]https://www.ll.mit.edu/ideval/data/1998data.html

First, we introduce some performance indicators: GPR is the true positive rate, FPR is the false positive rate, GCR is the correct classification rate, and ICR is the incorrect classification rate. All of these indicators will be calculated on unique elements that were not used while training.

The efficiency of the developed models was estimated with the help of a 5-fold cross-validation [40]. The dataset $\bar{\mathcal{X}}_C^{(TS)}$ containing $\bar{M}^*$=53,733 unique records was split into five disjoint subsets $\bar{\mathcal{X}}_C^{(TS)_1}, \ldots, \bar{\mathcal{X}}_C^{(TS)_5}$ for which $\#\bar{\mathcal{X}}_C^{(TS)_1} \approx \cdots \approx \#\bar{\mathcal{X}}_C^{(TS)_5}$. Furthermore, in each set $\bar{\mathcal{X}}_C^{(TS)_k}$ there are elements of all seven classes such that a subsample corresponding to each certain lth class possesses approximately an equal size inside each set $\bar{\mathcal{X}}_C^{(TS)_k}$: $\#\bar{\mathcal{X}}_{\{C_l\}}^{(TS)_1} \approx \cdots \approx \bar{\mathcal{X}}_{\{C_l\}}^{(TS)_5}$ $(k = 1, \ldots, 5, l = 0, \ldots, 6)$. The training and testing samples were taken in the ratio 3 : 2. The training process of basic classifiers was performed $3 \times \binom{5}{3} = 30$ times using the sets $\left\{\bar{\mathcal{X}}_C^{(TS)_{a_p}} \cup \bar{\mathcal{X}}_C^{(TS)_{b_p}} \cup \bar{\mathcal{X}}_C^{(TS)_{c_p}}\right\}_{p=1}^{3}$, where $a, b, c \in \mathbb{N} \wedge 1 \leqslant a < b < c \leqslant 5$. Depending on these sets, the testing set is constructed as $\left\{\bar{\mathcal{X}}_C^{(TS)_{d_p}} \cup \bar{\mathcal{X}}_C^{(TS)_{e_p}}\right\}_{p=1}^{3}$, where $d, e \in \mathbb{N} \setminus \{a, b, c\} \wedge 1 \leqslant d < e \leqslant 5$, and on each of these sets the values of indicators of the ith basic classifier $GPR_{i(de)_p}^{(BC)}, FPR_{i(de)_p}^{(BC)}, GCR_{i(de)_p}^{(BC)}, ICR_{i(de)_p}^{(BC)}$ $(i = 1, \ldots, 5)$ and jth aggregating composition $GPR_{j(de)_p}^{(AC)}, FPR_{j(de)_p}^{(AC)}, GCR_{j(de)_p}^{(AC)}, ICR_{j(de)_p}^{(AC)}$ $(j = 1, \ldots, 4)$. Such partitioning of the set $\bar{\mathcal{X}}_C^{(TS)}$ was performed three times $(p = 1, 2, 3)$, and every time the content of its ten subsets was randomly generated. The minimal, maximal, and average values of indicator GPR corresponding to the ith basic classifier and the jth aggregating composition can be defined as follows: $\frac{1}{10} \cdot \min\limits_{p=1,2,3} \sum\limits_{1 \leqslant d < e \leqslant 5} GPR_{i(de)_p}^{(BC)}, \frac{1}{10} \cdot \max\limits_{p=1,2,3} \sum\limits_{1 \leqslant d < e \leqslant 5} GPR_{i(de)_p}^{(BC)}, \frac{1}{30} \cdot \sum\limits_{p=1}^{3} \sum\limits_{1 \leqslant d < e \leqslant 5} GPR_{i(de)_p}^{(BC)}; \frac{1}{10} \cdot \min\limits_{p=1,2,3} \sum\limits_{1 \leqslant d < e \leqslant 5} GPR_{j(de)_p}^{(AC)}, \frac{1}{10} \cdot \max\limits_{p=1,2,3} \sum\limits_{1 \leqslant d < e \leqslant 5} GPR_{j(de)_p}^{(AC)}, \frac{1}{30} \cdot \sum\limits_{p=1}^{3} \sum\limits_{1 \leqslant d < e \leqslant 5} GPR_{j(de)_p}^{(AC)}$.
Other indicators can be calculated analogously.

5.6.2 Experiment 1

Figure 5.10 shows the performance indicators calculated using the 5-fold cross-validation for five basic classifiers and four aggregating compositions. These indicators correspond to the one-against-all scheme.

As a result of applying the Fix and Hodges method, the GCR indicator was increased by 0.707% on average in comparison with multilayer neural networks, which indicates the best result among the basic classifiers. The GPR indicator was increased slightly (by 0.021%) compared to the neuro-fuzzy networks.

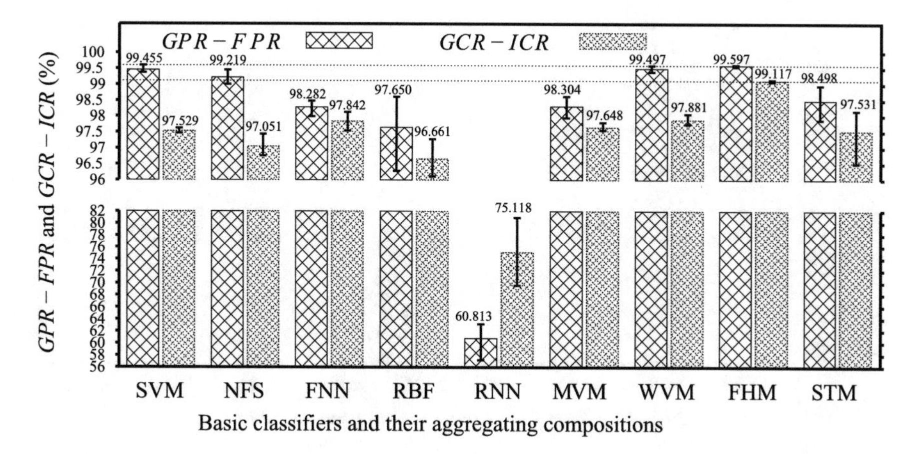

Fig. 5.10 Performance indicators for five basic classifiers and four aggregating compositions

The $GPR - FPR$ and $GCR - ICR$ indicators representing a trade-off between correct detection and false positives, and between correct classification and incorrect classification are depicted in Fig. 5.10. We denoted the amplitude of the change of these differences by a vertical line. Note that the greatest average value of the $GPR - FPR$ parameter belongs to the aggregating composition of the Fix and Hodges method, and the value of this parameter is 0.142% larger than in the case of support vector machines. At the same time, the $GCR - ICR$ indicator of the Fix and Hodges method is the largest, and exceeds the similar indicator of multilayer neural networks by 1.276%.

Within any model, an attempt to increase one indicator (true positive rate or correct classification rate), as a general rule, negatively affects other indicators (false positive rate or incorrect classification rate). However, as it was shown above, using aggregating compositions allows the exploitation of the benefits of strong models and eliminates the shortcomings of weak models. More precisely, it can be seen from Fig. 5.10 that the recurrent neural networks having the worst performance are not expected to make any positive contribution to the classifier ensemble. More detailed consideration of this effect is given in the next experiment.

5.6.3 Experiment 2

In this section we present results of experiments and evaluation of computational fault tolerance of aggregating compositions.

Figure 5.11 shows the performance indicators calculated using the 5-fold cross-validation and the introduction of two erroneous ("bad") classifiers.

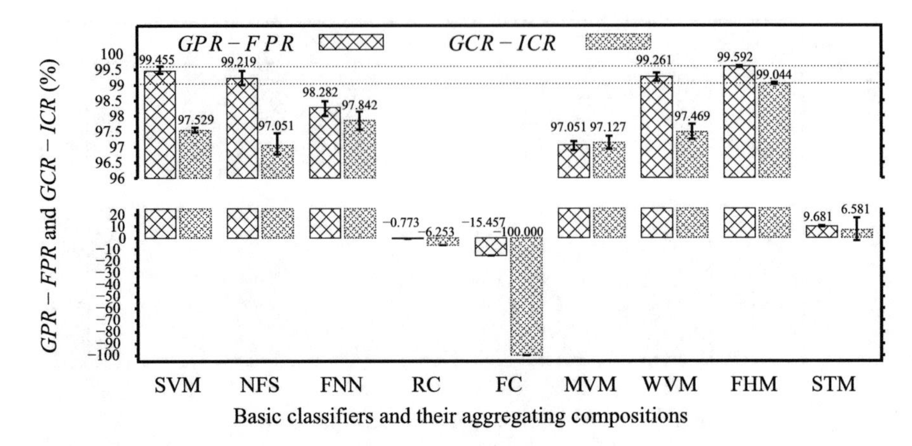

Fig. 5.11 Performance indicators for five basic classifiers and four aggregating compositions with the introduction of two erroneous classifiers

As earlier, these indicators correspond to the one-against-all scheme. The experiment described here is aimed at checking the properties of computational fault tolerance in relation to aggregating compositions.

In this experiment, the radial basis function networks and recurrent neural networks were replaced by two other classifiers: random and false classifiers. While the output of the first classifier is a random set of class labels, among which the correct label may exist, the second classifier deliberately excludes this possibility: its indicator of correct classification is always 0. The leading position in terms of GPR and GCR still belongs to the Fix and Hodges method. The values of these indicators exceed maximal values calculated for basic classifiers (neuro-fuzzy networks and multilayer neural networks) by 0.017% and 0.67%, respectively. Therefore, compared to the previous experiment, these values have fallen insignificantly (by $0.021 - 0.017 = 0.004\%$ and $0.707 - 0.67 = 0.037\%$). Similarly, we have calculated the loss caused by the introduction of "bad" classifiers for $GPR - FPR$ and $GCR - ICR$, which is $0.142 - 0.137 = 0.005\%$ and $1.276 - 1.203 = 0.073\%$, respectively. Based on this, it can be concluded that the proposed approach for combining the detectors using the Fix and Hodges method is computationally fault-tolerant. In the case of stacking, we have used a multilayer neural network as an aggregating composition. As one would expect, stacking is the most sensitive aggregating composition when introducing new classifiers. The average values of GPR and GCR have been decreased by 48.542 and 45.475%. This is due to the new classifiers (random and false classifiers) using the weights configured for radial basis function networks and recurrent neural networks.

5.7 Conclusion

In this chapter, models of binary classifiers applied to network attack detection have been presented. For optimizing the training process of such classifiers, a parallel genetic algorithm based on operators of crossover, mutation, and permutation has been introduced. For calculating the network parameters, we have described the sliding window method, which attempts to eliminate infrequent network bursts.

For constructing the multi-class models, several low-level schemes and aggregating compositions have been considered to combine binary classifiers. The experiments demonstrated the effectiveness of the proposed approach in terms of correct classification and true positive rates of network records.

Acknowledgements This research was supported by the Russian Science Foundation under grant number 18-11-00302.

References

1. Branitskiy A, Kotenko I (2016) Analysis and classification of methods for network attack detection. In: SPIIRAS Proceedings, vol 45(2), pp 207–244. https://doi.org/10.15622/sp.45.13 (in Russian)
2. Branitskiy A, Kotenko I (2015) Network attack detection based on combination of neural, immune and neuro-fuzzy classifiers. In: Plessl C, Baz DE, Cong G, Cardoso JMP, Veiga L, Rauber T (eds) Proceedings of the 18th IEEE International Conference on Computational Science and Engineering, IEEE Computer Society, Los Alamitos, CA, USA, pp 152–159. https://doi.org/10.1109/CSE.2015.26
3. Branitskiy A, Kotenko I (2017) Hybridization of computational intelligence methods for attack detection in computer networks. J Comput Sci 23:145–156. https://doi.org/10.1016/j.jocs.2016.07.010
4. Branitskiy A, Kotenko I (2017) Network anomaly detection based on an ensemble of adaptive binary classifiers. Computer Network Security. In: Rak J, Bay J, Kotenko I, Popyack L, Skormin V, Szczypiorski K (eds) Computer network security, pp 143–157. Springer, Cham. https://doi.org/10.1007/978-3-319-65127-9_12
5. Abraham A, Thomas J (2006) Distributed intrusion detection systems: a computational intelligence approach. In: Abbass HA, Essam D (eds) Applications of information systems to homeland security and defense. Idea Group, Hershey, PA, USA, pp 107–137. https://doi.org/10.4018/978-1-59140-640-2.ch005
6. Peddabachigari S, Abraham A, Grosan C, Thomas J (2007) Modeling intrusion detection system using hybrid intelligent systems. J Netw Comput Appl 30(1):114–132. https://doi.org/10.1016/j.jnca.2005.06.003
7. Mukkamala S, Sung AH, Abraham A (2003) Intrusion detection using ensemble of soft computing paradigms. In: Abraham A, Franke K, Köppen M (eds) Intelligent systems design and applications. Springer, Heidelberg, pp 239–248. https://doi.org/10.1007/978-3-540-44999-7_23
8. Mukkamala S, Sung AH, Abraham A (2005) Intrusion detection using an ensemble of intelligent paradigms. J Netw Comput Appl 28(2):167–182. https://doi.org/10.1016/j.jnca.2004.01.003
9. Toosi AN, Kahani M (2007) A new approach to intrusion detection based on an evolutionary soft computing model using neuro-fuzzy classifiers. Comput Commun 30(10):2201–2212. https://doi.org/10.1016/j.comcom.2007.05.002
10. Amini M, Rezaeenour J, Hadavandi E (2014) Effective intrusion detection with a neural network ensemble using fuzzy clustering and stacking combination method. J Comput Sec 1(4):293–305

11. Wang G, Hao J, Ma J, Huang L (2010) A new approach to intrusion detection using artificial neural networks and fuzzy clustering. Expert Syst Appl 37(9):6225–6232. https://doi.org/10.1016/j.eswa.2010.02.102
12. Chandrasekhar AM, Raghuveer K (2013) Intrusion detection technique by using k-means, fuzzy neural network and SVM classifiers. In: Proceedings of the 2013 International Conference on Computer Communication and Informatics. Curran Associates, Red Hook, NY, USA. https://doi.org/10.1109/ICCCI.2013.6466310
13. Saied A, Overill RE, Radzik T (2016) Detection of known and unknown DDoS attacks using artificial neural networks. Neurocomputing 172:385–393. https://doi.org/10.1016/j.neucom.2015.04.101
14. Agarwal B, Mittal N (2012) Hybrid approach for detection of anomaly network traffic using data mining techniques. Proc Tech 6:996–1003. https://doi.org/10.1016/j.protcy.2012.10.121
15. He H-T, Luo X-N, Liu B-L (2005) Detecting anomalous network traffic with combined fuzzy-based approaches. In: Huang D-S, Zhang X-P, Huang G-B (eds) Advances in intelligent computing. Springer, Heidelberg, pp. 433–442. https://doi.org/10.1007/11538356_45
16. Kolmogorov AN (1957) On the representation of continuous functions of several variables as superpositions of continuous functions of one variable and addition. In: Tikhomirov VM (ed) Selected works of A. N. Kolmogorov, pp. 383–387. https://doi.org/10.1007/978-94-011-3030-1_56
17. Cybenko G (1989) Approximation by superpositions of a sigmoidal function. Math Control Signal 2(4):303–314. https://doi.org/10.1007/BF02551274
18. Hornik K, Stinchcombe M, White H (1989) Multilayer feedforward networks are universal approximators. Neural Netw 2(5):359–366. https://doi.org/10.1016/0893-6080(89)90020-8
19. Funahashi K-I (1989) On the approximate realization of continuous mappings by neural networks. Neural Netw 2(3):183–192. https://doi.org/10.1016/0893-6080(89)90003-8
20. Haykin SS (2011) Neural networks and learning machines, 3rd edn. Pearson, Upper Saddle River, NJ, USA
21. Riedmiller M, Braun H (1993) A direct adaptive method for faster backpropagation learning: the RPROP algorithm. In: Proceedings of IEEE International Conference on Neural Networks, vol 1. IEEE, New York, pp 586–591. https://doi.org/10.1109/ICNN.1993.298623
22. Fahlman SE (1988) Faster-learning variations on back-propagation: an empirical study. In: Proceedings of the 1988 connectionist models summer school. Morgan Kaufmann, San Francisco, pp 38–51
23. Levenberg K (1944) A method for the solution of certain non-linear problems in least squares. Q Appl Math 2(2):164–168. https://doi.org/10.1090/qam/10666
24. Marquardt DW (1963) An algorithm for least-squares estimation of nonlinear parameters. J Soc Ind Appl Math 11(2):431–441. https://doi.org/10.1137/0111030
25. Jordan ML (1986) Attractor dynamics and parallelism in a connectionist sequential machine. In: Proceedings of the eighth annual conference of the cognitive science society. Lawrence Erlbaum Associates, Hillsdale, NJ, USA, pp 531–546
26. Takagi T, Sugeno M (1985) Fuzzy identification of systems and its applications to modeling and control. IEEE T Syst Man Cyb SMC-15(1):116–132. https://doi.org/10.1109/TSMC.1985.6313399
27. Jang J-SR (1993) ANFIS: adaptive-network-based fuzzy inference system. IEEE T Syst Man Cyb 23(3):665–685. https://doi.org/10.1109/21.256541
28. Strang G (2016) Introduction to linear algebra, 5th edn. Cambridge Press, Wellesley, MA, USA
29. Vapnik V (1995) The nature of statistical learning theory. Springer-Verlag, New York. https://doi.org/10.1007/978-1-4757-2440-0
30. Hsu CW, Lin CJ (2002) A comparison of methods for multiclass support vector machines. IEEE T Neural Networ 13(2):415–425. https://doi.org/10.1109/72.991427
31. Drucker H, Burges CJC, Kaufman L, Smola A, Vapnik V (1997) Support vector regression machines. Advances in neural information processing systems 9. MIT Press, Cambridge, MA, USA, pp 155–161

32. Müller KR, Smola AJ, Rätsch G, Schölkopf B, Kohlmorgen J, Vapnik V (1997) Predicting time series with support vector machines. In: Gerstner W, Germond A, Hasler M, Nicoud J-D (eds) Artificial neural networks – ICANN'97, pp 999–1004. https://doi.org/10.1007/BFb0020283
33. Kuhn HW, Tucker AW (1951) Nonlinear programming. In: Neyman J (ed) Proceedings of 2nd Berkeley Symposium on Mathematical Statistics and Probabilistics. University of California Press, Berkeley, CA, USA, pp 481–492
34. Platt J (1998) Sequential minimal optimization: a fast algorithm for training support vector machines (1998). https://www.microsoft.com/en-us/research/publication/sequential-minimal-optimization-a-fast-algorithm-for-training-support-vector-machines
35. Shawe-Taylor J, Cristianini N (2004) Kernel methods for pattern analysis. Cambridge University Press, New York
36. Jolliffe IT (2011) Principal component analysis. In: Lovric M (ed) International encyclopedia of statistical science. Springer, Heidelberg. https://doi.org/10.1007/978-3-642-04898-2_455
37. Fix E, Hodges J (1951) Discriminatory analysis. Nonparametric discrimination: consistency properties. Technical Report 4, USAF School of Aviation Medicine, Randolph Field, TX, USA
38. McHugh J (2000) Testing intrusion detection systems: a critique of the 1998 and 1999 DARPA intrusion detection system evaluations as performed by Lincoln laboratory. ACM T Inform Syst Se 3(4):262–294. https://doi.org/10.1145/382912.382923
39. Mahoney MV, Chan PK (2003) An analysis of the 1999 DARPA/Lincoln Laboratory evaluation data for network anomaly detection. In: Vigna G, Kruegel C, Jonsson E (eds) Recent advances in intrusion detection. Springer, Heidelberg, pp 220–237. https://doi.org/10.1007/978-3-540-45248-5_13
40. Refaeilzadeh P, Tang L, Liu H (2009) Cross-validation. In: Liu L, Özsu MT (eds) Encyclopedia of database systems. Springer, Boston, MA, USA. https://doi.org/10.1007/978-0-387-39940-9_565

Chapter 6
Machine Learning Algorithms for Network Intrusion Detection

Jie Li, Yanpeng Qu, Fei Chao, Hubert P. H. Shum, Edmond S. L. Ho, and Longzhi Yang

Abstract Network intrusion is a growing threat with potentially severe impacts, which can be damaging in multiple ways to network infrastructures and digital/intellectual assets in the cyberspace. The approach most commonly employed to combat network intrusion is the development of attack detection systems via machine learning and data mining techniques. These systems can identify and disconnect malicious network traffic, thereby helping to protect networks. This chapter systematically reviews two groups of common intrusion detection systems using fuzzy logic and artificial neural networks, and evaluates them by utilizing the widely used KDD 99 benchmark dataset. Based on the findings, the key challenges and opportunities in addressing cyberattacks using artificial intelligence techniques are summarized and future work suggested.

6.1 Introduction

Cybersecurity can be assisted by a set of techniques that protect cyberspace and ensure the integrity, confidentiality, and availability of networks, applications, and data. Cybersecurity techniques also have the potential to defend against and recover from any type of attack. More devices, namely, Internet of Things (IoT) devices, are connected to the cyberspace, and cybersecurity has become an elevated concern affecting governments, businesses, other organizations, and individuals. The scope of cybersecurity is broad, and can be grouped into five areas: critical infrastructure, network security, cloud security, application security, and IoT security. Network

---

J. Li · H. P. H. Shum · E. S. L. Ho · L. Yang (✉)
Northumbria University, Newcastle upon Tyne, UK
e-mail: longzhi.yang@northumbria.ac.uk

Y. Qu
Dalian Maritime University, Dalian, People's Republic of China
e-mail: yanpengqu@dlmu.edu.cn

F. Chao
Xiamen University, Xiamen, People's Republic of China
e-mail: fchao@xmu.edu.cn

© Springer Nature Switzerland AG 2019

L. F. Sikos (ed.), *AI in Cybersecurity*, Intelligent Systems Reference Library 151,
https://doi.org/10.1007/978-3-319-98842-9_6

security is an important challenge in the field of cybersecurity, because networks provide the means for the crucial access to others devices, and for connectivity between all the assets in cyberspace. Severe network attacks can lead to system damage, network paralysis, and data loss or leakage. Network intrusion detection systems (NIDS) attempt to identify unauthorized, illicit, and anomalous behavior based solely on network traffic to support decision-making in network preventative actions by network administrators.

Traditional network intrusion detection systems are mainly developed using available knowledge bases, which are comprised of the specific patterns or strings that correspond to already known network behaviors, i.e., normal traffic and abnormal traffic [1]. These patterns are used to check monitored network traffic to recognize possible threats. Typically, the knowledge bases of such systems are defined based on expert knowledge, and the patterns must be updated to ensure the coverage of new threats [2]. Therefore, the detection performance of traditional network intrusion detection systems depends highly on the quality of the knowledge base. From the theoretical point of view, network intrusion detection systems mainly aim to classify the monitored traffic as either "legitimate" or "malicious." Therefore, machine learning approaches are appropriate to solve such problems; and they have recently been widely applied to help better manage network intrusion detection issues.

Machine learning (ML) is a field of artificial intelligence, which refers to a set of techniques that give computer systems the ability to "learn." Typically, machine learning algorithms, such as artificial neural networks, learn from data samples to categorize or find patterns in the data, and enable computer systems to make predictions on new or unseen data instances based on the discovered patterns [3]. Depending on the way of learning, machine learning can be further grouped into two main categories: supervised learning and unsupervised learning. Supervised learning discovers the patterns to map an input to an output based on the labeled input-output pairs of data samples [4]. The classification problem is a typical supervised learning problem, which has been commonly used for solving NIDS problems, such as those reported in [5–8]. The goal of unsupervised learning is to find a mapping that is able to describe a hidden structure from unlabeled data samples. It is a powerful tool for identifying structures when unlabeled data samples are given [4]. Thanks to the relaxation of the requirement for labels of training data in unsupervised learning, various unsupervised learning approaches have also been widely applied for NIDS problems, such as the clustering-based NIDS [9] and self-organizing map-based NIDS [10].

This chapter focuses primarily on network intrusion detection systems, and particularly how the machine learning and data mining techniques can help in developing network intrusion detection systems. The chapter first systematically reviews intrusion detection techniques from the perspective of both hardware deployment and software implementation. The two most commonly used NIDS development methods and the three most commonly used detection methodologies are reviewed; these are followed by the investigation of applying machine learning and data mining techniques in the implementation of intrusion detection systems. Two representative machine learning approaches, including fuzzy inference systems and artificial neural networks, are of particular interest in this chapter, because they are the machine learning and data mining techniques most suitable for supporting intrusion detection systems. Traditionally, fuzzy inference systems are not classified as machine learning algorithms, however, the rule base generation mechanism follows the data mining principle; therefore, fuzzy inference systems with automatic rule base generation can also be considered machine learning. Finally, the intrusion detection systems developed upon these machine learning approaches are evaluated using the widely used KDD 99 benchmark dataset.

The remainder of this chapter is organized as follows. Section 6.2 introduces the hardware deployment methods of network intrusion detection systems and detection methodologies. Section 6.3 reviews the existing machine learning-based network intrusion detection systems using fuzzy inference systems and artificial neural networks. The limitations and potential solutions of both techniques are also discussed in this section. Section 6.4 evaluates the studied systems using the well-known benchmark dataset KDD 99. Section 6.5 concludes the chapter and sets directions for future work.

6.2 Network Intrusion Detection Systems

Network intrusion detection systems are software-based or hardware-based tools that are used to monitor network traffic, i.e., to analyze them for signs of possible attacks or suspicious activities. Usually one or more network traffic sensors are used to monitor network activity on one or more network segments. The system constantly performs analysis and watches for certain patterns of passing traffic in a monitored network environment. If the detected traffic patterns match the defined signatures or policies in the knowledge base (e.g., based on a fuzzy rule base or a trained neural network), a security alert is generated.

6.2.1 Deployment Methods

There are multiple methods that can be adopted to deploy a NIDS in order to capture and monitor traffic in a network environment, with passive deployment and in-line deployment being the most commonly used, as shown in Fig. 6.1a and b.

In the passive deployment method, the NIDS device is connected to a network switch, which is deployed between the main firewall and the internal network. The switch is usually configured with a port mirroring technology, such as the Mirror Port supported by HP and the Switched Port Analyzer (SPAN) supported by Cisco. These port mirroring technologies are able to copy all network traffic, including incoming and outgoing traffic, to a particular interface of the NIDS for the purpose of traffic monitoring and analysis. This method usually requires a high-end network switch in order to enable the port mirroring technologies. There is a special case of passive deployment, which is the passive network TAP (Terminal Access Point) [11].

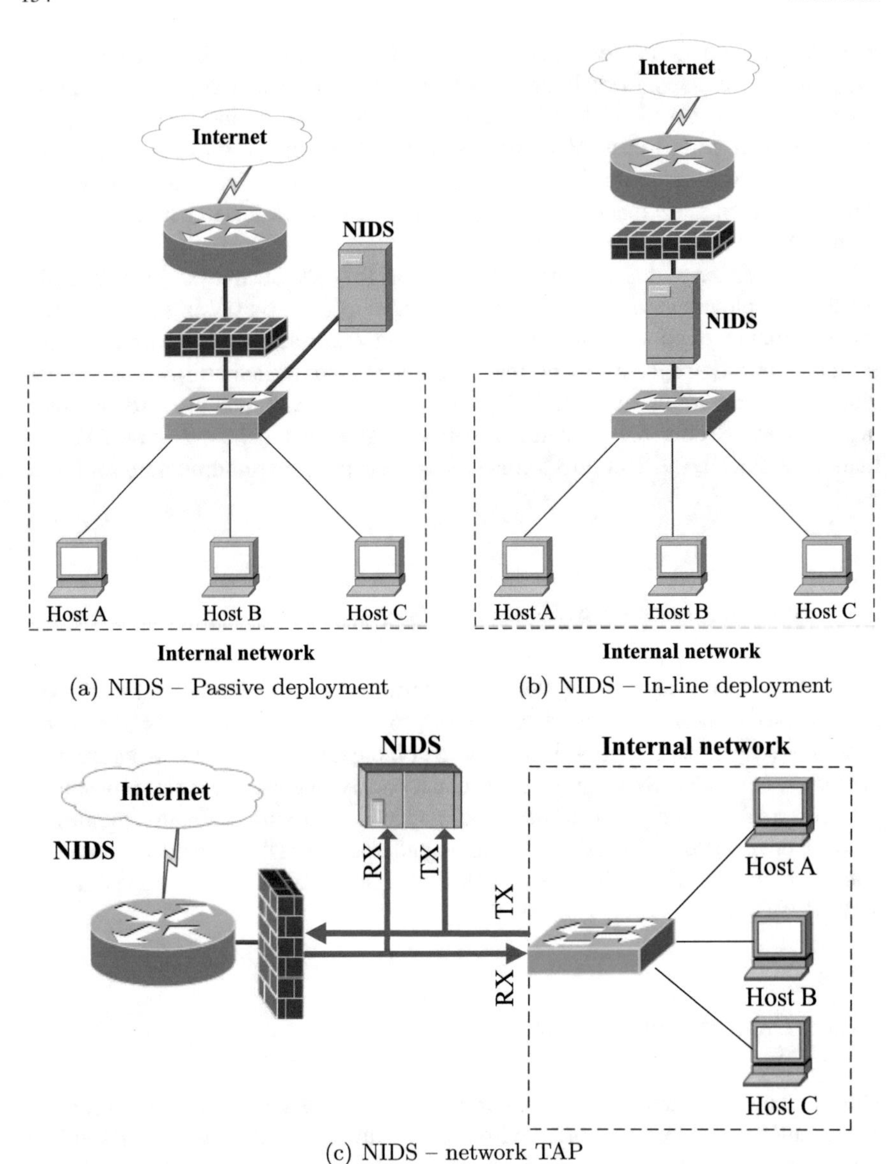

Fig. 6.1 Deployment methods for intrusion detection systems

In particular, a network TAP uses pairs of cables included in the original Ethernet cable, as illustrated in Fig. 6.1c, to send a copy of the original network traffic to the NIDS.

The in-line deployment method deploys NIDS devices the same way as firewalls, which allows all traffic to pass directly through the NIDS. Therefore, this deployment method does not require any particularly high-end network device, which is an ideal solution for those environments in which port mirroring technologies are unavailable, such as a small branch office with low-end networking equipment.

It is important to note that the deployment methods should be carefully selected while taking into account the network topology for optimal performance. For instance, in the example shown in Fig. 6.1a, the port mirroring method is not only able to monitor the outgoing traffic between the internal network and the Internet, but also the internal traffic between hosts A, B, and C. However, the network TAP and in-line deployment method are only able to monitor the outgoing traffic that is generated between the internal network and the Internet. Therefore, the NIDS, which is deployed by either the network TAP or the in-line method, will not notice if there is suspicious traffic between two client machines. In addition, because the port mirroring method uses a signal network interface to monitor the entire switch traffic, traffic congestion may occur if the switch backbone traffic is beyond the capacity of the bandwidth of the monitored port. Therefore, it is a good strategy to deploy multiple NIDSes in complex network environments so that these blind spots can be eliminated.

6.2.2 Detection Methodologies

Generally speaking, intrusion detection methodologies can be grouped into three major categories: signature-based detection, anomaly-based detection, and specification-based detection [12].

The signature-based NIDS, also called knowledge-based detection or misuse detection, refers to the detection of attacks or threats by looking for specific patterns or strings that correspond to already known attacks or threats. These specific patterns or strings are saved in a knowledge base, such as the byte sequences of the network traffic, known malicious instruction sequences exploited by malware, the specific ports a host tries to access, etc. Signature-based detection is a process that compares known patterns against monitored network traffic to recognize possible intrusions. Therefore, signature-based detection is able to effectively detect known threats in a network environment, and its knowledge bases are usually generated by experts. A good example for this type of detection is a large amount of failed login attempts that have been detected in a Telnet session.

Anomaly-based detection primarily focuses on normal traffic behaviors rather than specific attack behaviors, which overcomes the limitation of signature-based detection that is only able to detect known attacks. This method is usually comprised of two processes: a training process and a detection process. In the training phase, machine learning algorithms are usually adopted to develop a model of trustworthy activity based on the behavior of the network traffic without attacks. In the detection phase, the developed trustworthy activity model is compared to the currently monitored traffic behavior, and any deviations indicate a potential threat. The anomaly-based detection method is usually adopted to detect unknown attacks

[13–18]. However, the effectiveness of anomaly-based detection is greatly affected by the selected features the machine learning algorithms use. Unfortunately, the selection of the appropriate set of features has proven to be a big challenge. Also, the observed systems' behavior constantly changes, which causes anomaly-based detection to produce a weak profile accuracy.

Specification-based detection is similar to the anomaly-based detection method as it also detects attacks as deviations from normal behavior. However, specification-based approaches are based on manually developed specifications that characterize legitimate behaviors rather than relying on machine learning algorithms. Although this method is not characterized by the high rate of false alarms typical to anomaly-based detection methods, the development of detailed specifications can be time-consuming. Because they detect attacks as deviations from legitimate behaviors, specification-based approaches are commonly used for unknown attack detection [19, 20]. In addition, multiple detection methodologies can be adopted jointly to provide a more extensive and accurate detection [21].

6.3 Machine Learning in Network Intrusion Detection

Machine learning and data mining techniques work by establishing an explicit or implicit model that enables the analyzed patterns to be categorized. In general, machine learning techniques are able to deal with three common problems: classification, regression, and clustering. Network intrusion detection can be considered as a typical classification problem. Therefore, a labeled training dataset is usually required for system modeling. A number of machine learning approaches have been used to solve network intrusion detection problems, and all of them consist of three general phases (as illustrated in Fig. 6.2):

- *Preprocessing*: the data instances that are collected from the network environment are structured, which can then be directly fed into the machine learning algorithm. The processes of feature extraction and feature selection are also applied in this phase.
- *Training*: a machine learning algorithm is adopted to characterize the patterns of various types of data, and build a corresponding system model.
- *Detection*: once the system model is built, the monitored traffic data will be used as system input to be compared to the generated system model. If the pattern of the observation is matched with an existing threat, an alarm will be triggered.

Both supervised and unsupervised machine learning approaches have already been utilized to solve network intrusion detection problems. For instance, supervised learning-based classifiers have been successfully employed to detect unauthorized access, such as k-nearest neighbor (k-NN) [6], support vector machine (SVM) [22], decision tree [23], naïve Bayes network [7], random forests [5], and artificial neural networks (ANN) [24]. In addition, unsupervised learning algorithms, including k-means clustering [25] and self-organized maps (SOM) [10], have also been applied

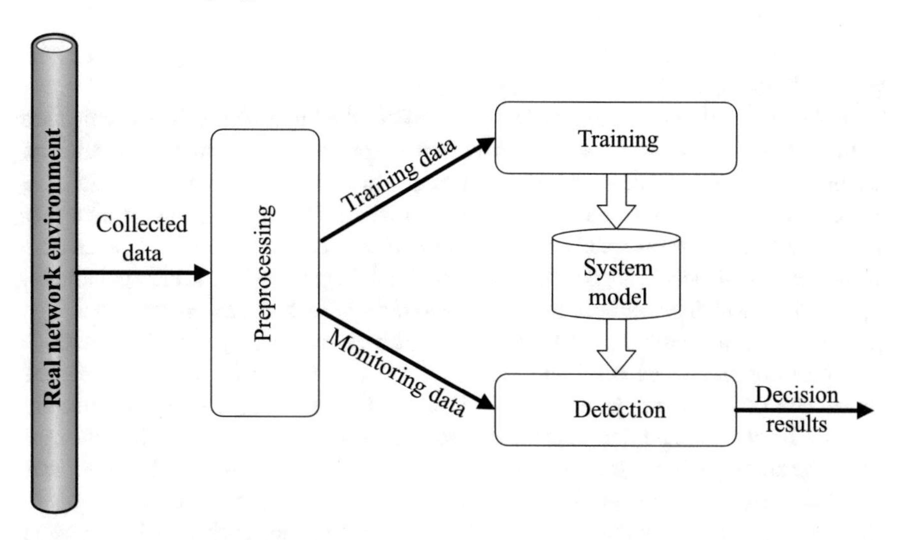

Fig. 6.2 ML-NIDS architecture

to deal with network intrusion detection problems, with good results. For various reasons, such as the imbalance of training datasets and the high cost of computational requirement, it is currently very difficult to design a single machine learning approach that outperforms the existing ones. Therefore, hybrid machine learning approaches, such as clustering with classifier [16, 26] and hierarchical classifiers [27], have attracted a lot of attention in recent years. In addition, some data mining approaches have also been successfully utilized to solve intrusion detection problems. For instance, data mining approaches are employed to generate a fuzzy rule base, and a fuzzy inference approach is then applied for threat detection [14]. This section examines the existing NIDSes utilizing two approaches, namely, fuzzy inference systems and artificial neural networks.

6.3.1 Fuzzy Inference Systems

Due to their great ability to deal with uncertainty, fuzzy inference systems (FIS) have been widely used for detecting potential network threats. Generally speaking, fuzzy inference systems are built upon fuzzy logic to map system inputs and outputs. A typical fuzzy inference system consists of two main parts: a rule base (or knowledge base) and an inference engine. A number of inference engines are well established, with the Mamdani inference [28] and the TSK inference [29] being the most widely used. Although fuzzy sets are used in both rule antecedents and rule consequences by the Mamdani fuzzy model, which is more intuitive and suitable for handling linguistic variables, a defuzzification progress is required to transfer the fuzzy outputs to crisp outputs. In contrast, the TSK inference approach produces crisp outputs directly, as crisp polynomials are used as rule consequences.

For a fuzzy inference-based NIDS (FIS-NIDS), the important features, which are extracted from the network packets, are used in the pre-detector component to analyze events with the set of rules to determine whether any incoming events have intrusive patterns or not. The set of rules is called a fuzzy rule base, which can be either predefined by expert knowledge (knowledge-driven), or extracted from labeled data instances (data-driven) [30, 31]. In contrast to knowledge-driven rule base generation approaches, which essentially limit the system's applicability as expert knowledge is not always available in some areas, data-driven rule base generation methods are most commonly used for intelligent NIDSes. Several data-driven approaches have been proposed to generate a rule base for FIS-NIDS use, which are usually derived from complete and dense datasets [32, 33]. The generated rule bases are often optimized using a general optimization technique, such as genetic algorithms (GA), for optimal system performance. As the used datasets are dense and complete, the resulted rule bases are generally dense and complete, each of which covers the entire input domain, and accordingly the resulted fuzzy models often yield to great reasoning performance. However, these systems will suffer if only incomplete, imbalanced, and sparse datasets are available. In addition, these systems are usually signature-based NIDSes, which are only able to detect known network threats for which the intrusive patterns have been covered in the rule base.

In order to address the previous limitations, fuzzy interpolation has been used to develop NIDSes [18, 34]. Briefly, fuzzy interpolation enhances conventional fuzzy inference systems to work with sparse fuzzy rule bases, by which some inputs or observations are not covered [35]. Using fuzzy

interpolation techniques, even if the traffic patterns of the incoming event do not match with any of the patterns stored in the rule base, an approximated result can still be obtained by considering the similar patters expressed as rules in the current rule base. A number of fuzzy interpolation approaches have been proposed in literature [36–47], many of which have already been applied to solve real-world problems [48–51].

A data-driven fuzzy interpolation-based NIDS can be developed in four steps: 1) training dataset generation and preprocessing, 2) rule base initialization, 3) rule base optimization, and 4) intrusion detection by fuzzy interpolation [14, 52], as illustrated in Fig. 6.3. These key steps are detailed in the following sections.

6.3.1.1 Dataset Generation and Preprocessing

The training dataset can either be collected from a real-world network environment, or it can be developed from an existing dataset. Whichever method is selected, the important features, which are selected for system modeling, have to be identified. In general, a number of features can be monitored by networking tools for network analysis during data packet transmission over the network, but some of these features are redundant or noisy. Therefore, a well-thought manual feature selection process

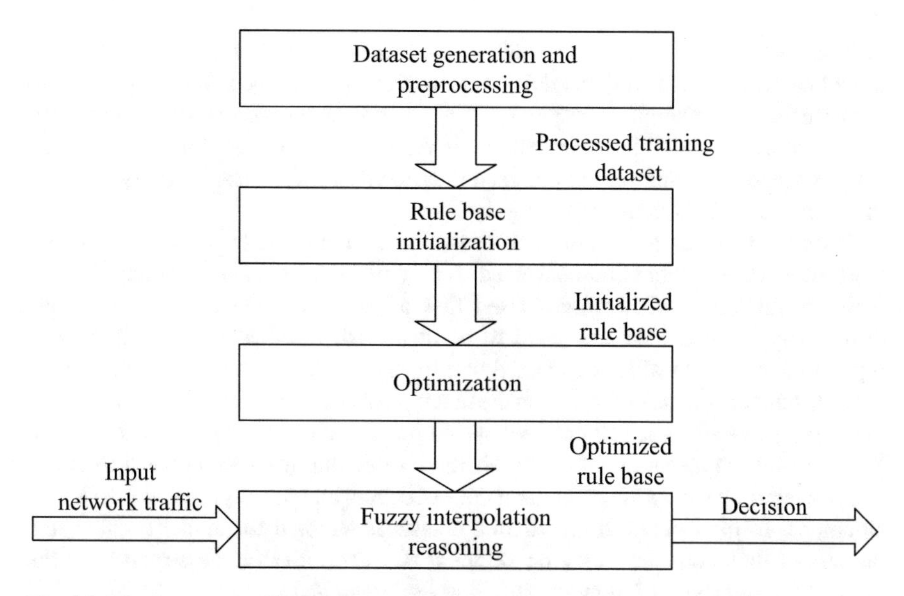

Fig. 6.3 The framework of TSK– based NIDS

is often required for network attack detection [53]. This common practice is also applied here. In particular, four important features identified by experts are selected as NIDS signature for the proposed FIS-NIDS, which are listed in Table 6.1.

The establishment of the optimal number of features that should be retained in datasets by feature selection methods is always an argued point, because feature selection usually causes information loss from the original dataset. Several pieces of work in the area of feature selection have claimed that more attributes generally lead to better approximations [54–57]. This can be the case for perfect, entirely consistent, and noise-free data, with all features being independent. Generally speaking, feature relevancy and redundancy have to be considered by feature selection methods before the application of machine learning approaches [58, 59]. The selected features should be highly relevant to the problem and non-redundant to be useful and efficient [60]. In fact, a large volume of published results in relevant literature has

Table 6.1 Features used in the NIDS

Feature	Description
Source bytes	The number of data bytes sent by the source IP host
Destination bytes	The number of data bytes sent by the destination IP host
Count	The number of connections to the same host as the current connection in the past 2 seconds
Dst_Host_Diff_Rate	% of connections whose ports are different, among the past 100 connections with the same destination IP

demonstrated that smaller number of selected features can lead to much-improved modeling accuracy [61–66]. In addition, more attributes retained in datasets will also increase the computational complexity [60]. Therefore, it is necessary to consider as many features as possible under certain circumstances especially for noise-free and fully consistent datasets, but in others, a minimal subset of features satisfying some predefined criteria is more appealing.

Once the features are determined for machine learning, datasets for a given network of a particular environment need to be collected for model training. This is typically implemented in stages based first on an attack-free network, and then different types of attacks that need to be identified. In other words, data regarding normal network traffic is collected first from a threat-free network environment. Then, a number of attacks simulating the first type of attack are artificially launched so that this type of attack is sufficiently covered by the dataset. This process is repeated for every other type of attacks until all the classes that need to be considered are fully covered by the dataset. The final dataset covers all attack types and attack-free situations. In most cases, if an existing dataset is adopted for model training, the process of data collection may be skipped. However, ideally, the structure of the existing dataset should follow the structure explained above.

6.3.1.2 Rule Base Initialization

Suppose that the training dataset (T) contains l ($l \geqslant 1, l \in \mathbb{N}$) labeled classes, which covers $l - 1$ types of attacks and the normal situation. As illustrated in Fig. 6.4, the system first divides the training dataset T into l sub-datasets $T_1, T_2, \ldots, T_l$, each representing a type of attack or the normal traffic (i.e., $T = \cup_{s=1}^{l} T_s$).

Then, the K-means, one of the most widely used clustering algorithms, is employed to each sub-dataset to group the data points into k clusters based on their feature values. Note that the value of k in the K-means algorithm has to be predefined to enable the application of the algorithm. The Elbow method [67], which determines the number of clusters based on the criteria that adding another cluster is not much better for modeling the dataset, has been employed for determining the value of k. Based on this, each determined cluster is expressed as a fuzzy rule that contributes to the TSK rule base.

In this work, a 0-order TSK fuzzy model is adopted. All data instances in each class share the class label (an integer number), which is utilized as the consequent of the corresponding TSK rule. The triangular membership function is utilized in the rule antecedents. The support of the triangular fuzzy set is expressed as the span of the cluster along this input dimension, and the core of the corresponding fuzzy set is set as the cluster center. The final TSK fuzzy rule base is generated by combining all the extracted rules from all l sub-datasets, which is illustrated as follows:

$$
\begin{aligned}
R_{t_s}^{s} :\ & \textbf{IF}\ x_1 \text{ is } A_1^{st_s} \text{ and } x_2 \text{ is } A_2^{st_s} \text{ and } x_3 \text{ is } A_3^{st_s} \text{ and } \text{ and } x_4 \text{ is } A_4^{st_s}, \\
& \textbf{THEN}\ z = s,
\end{aligned}
\tag{6.1}
$$

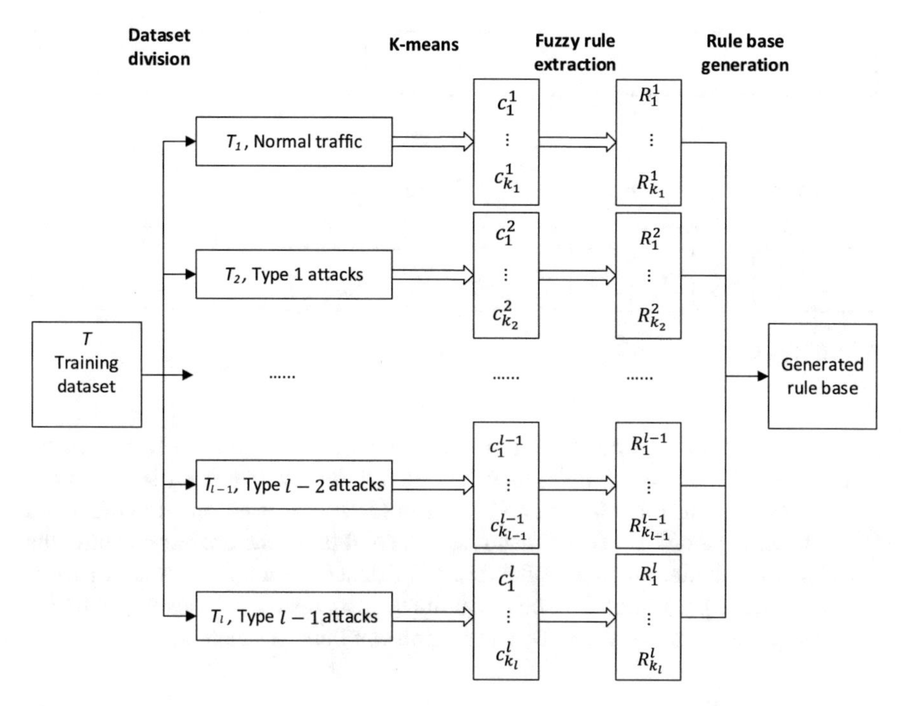

Fig. 6.4 Rule base generation

where $s = \{1, \ldots, l\}$ represents the sth sub dataset that indicates the sth type of network traffic, $t_s = \{1, \ldots, k_s\}$ denotes the tth cluster in the sth sub-dataset. The number of rules in this rule base is equal to the sum of the numbers of clusters for all the sub-datasets (i.e., $k_1 + k_2 + \cdots + k_l$).

6.3.1.3 Rule Base Optimization

The generated initial rule base can be employed for intrusion detection, but with relatively poor performance. In order to increase the detection performance, a genetic algorithm (GA) is adopted here to fine-tune the membership functions involved in the initial rule base. Assume that a given initial TSK rule base is comprised of n fuzzy rules of the form shown in Eq. 6.1. Suppose a chromosome, denoted as I, is used to represent a potential solution in the GA, which is coded to represent the parameters of all rules in the rule base, as shown in Fig. 6.5. Based on this, the initial population $\mathbb{P} = \{I_1, I_2, \ldots, I_{|\mathbb{P}|}\}$ can be formed by taking the parameters of the initial rule base and its random variations. During the optimization process, the number of chromosomes is selected for offspring reproduction by applying the genetic operators of crossover and mutation. Specifically, the *fitness proportionate selection method*, also known as the *roulette wheel selection*, is implemented in this work for chromosome

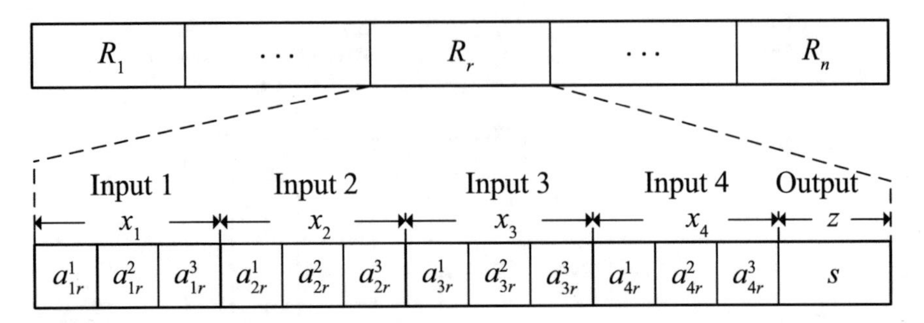

Fig. 6.5 Chromosome encoding

selection, and the signal point crossover and mutation operators are employed for reproduction. In addition, in order to make sure that the resultant fuzzy sets are valid and convex, the constraint $a^1_{ir} < a^2_{ir} < a^3_{ir}$, $i = \{1, 2, 3, 4\}$ is enforced to the genes during optimization. The selection and reproduction processes are iterated until the predefined maximum number of iterations is reached, or until the system performance reaches a predefined threshold. Optimized parameters and thus an optimized rule base can be obtained when the termination condition is satisfied.

6.3.1.4 Intrusion Detection by TSK-Interpolation:

Once the rule base is generated, the TSK+ fuzzy inference approach can be deployed to perform inferences for attack detection. In order to generate network intrusion alerts in real time, the system is deployed by one of the deployment methods introduced in Sect. 6.2.1, which keeps capturing network traffic data for analysis. For each captured network packet, four important features, detailed in Table 6.1, are extracted and fed into the proposed system. From this input, the TSK+ fuzzy inference approach will classify the types of network traffic using the generated rule base. Assume that an optimized TSK fuzzy rule base is comprised of n rules as follows:

$$\begin{aligned} R_1 &: \textbf{IF}\, x_1 \text{ is } A^1_1 \text{ and } x_2 \text{ is } A^1_2 \text{ and } x_3 \text{ is } A^1_3 \text{ and } x_4 \text{ is } A^1_4 \textbf{ THEN}\, z = \mathbb{Z}_1, \\ &\dots \\ R_n &: \textbf{IF}\, x_1 \text{ is } A^n_1 \text{ and } x_2 \text{ is } A^n_2 \text{ and } x_3 \text{ is } A^n_3 \text{ and } x_4 \text{ is } A^n_4 \textbf{ THEN}\, z = \mathbb{Z}_n, \end{aligned} \tag{6.2}$$

where $A^i_k (k \in \{1, 2, 3, 4\}$ and $i \in \{1, \dots, n\})$ represents a normal and convex triangular fuzzy set in the rule antecedent denoted accordingly as $(a^i_{k_1}, a^i_{k_2}, a^i_{k_3})$, and $\mathbb{Z}_i$ is an integer number that indicates the type of network traffic, whether it is normal traffic or a particular type of attack. By taking a captured network packet as an example, the working procedure of the TSK+ fuzzy inference for intrusion detection can be summarized as the following steps:

1. Extract the four feature values from the network packet, and express them in the form $I = \{x_1^*, x_2^*, x_3^*, x_4^*\}$, which will be used as the system input. Note that the extracted feature values are normally crisp values. They have to be represented as fuzzy sets of the form $A_k^* = (x_k^*, x_k^*, x_k^*)$, where $k = \{1, 2, 3, 4\}$, for future use.

2. Determine the matching degree $S(A_k^*, A_k^i)$ between the inputs $I = \{A_1^*, A_2^*, A_3^*, A_4^*\}$ and rule antecedents $(A_1^i, A_2^i, A_3^i, A_4^i)$ for each rule R_i, $i = \{1, \ldots, n\}$ using:

$$S(A_k^*, A_k^i) = \left(1 - \frac{\sum_{j=1}^{3} |x_k^* - a_{kj}^i|}{3} \right) \cdot DF \ , \tag{6.3}$$

where DF, termed as distance factor, is a function of the distance between the two fuzzy sets of interest, which is defined as follows:

$$DF = 1 - \frac{1}{1 + e^{-sd+5}} \ , \tag{6.4}$$

where s ($s > 0$) is a sensitivity factor, and d represents the Euclidean distance between the two fuzzy sets. A smaller s value results in a similarity degree more sensitive to the distance of the two fuzzy sets.

3. Obtain the firing degree of each rule by integrating the matching degrees of its antecedents and the given input values as follows:

$$\alpha_i = S(A_1^*, A_1^i) \wedge S(A_2^*, A_2^i) \wedge S(A_3^*, A_3^i) \wedge S(A_4^*, A_4^i) \ , \tag{6.5}$$

where $\wedge$ is a t-norm usually implemented as a minimum operator.

4. Integrate the sub-consequences from all rules to get the final output using the following formula:

$$z = \frac{\sum_{i=1}^{n} \alpha_i \cdot \mathbb{Z}_i}{\sum_{i=1}^{n} \alpha_i} . \tag{6.6}$$

5. Apply the round function on the final output to obtain the integer number that indicates the network traffic type for the given network packet.

As discussed above, if an unknown network's threat behavior or traffic pattern has been captured, a result of "network security alert" can still be expected by considering all fuzzy rules in the rule base.

6.3.2 Artificial Neural Networks

An artificial neural network (ANN) is an information processing system inspired by biological nervous systems that constitute animal brains, which is one of the most widely used machine learning algorithms [68]. Typically, an ANN is composed of two main parts: a set of simple processing units, also known as nodes or artificial neurons, and the connections between these. These simple units or nodes are organized in layers, which usually consist of the input, output, and hidden layers. The hidden layers are those between the input and the output layers. Once the set of processing units and their connections are determined, or an ANN is built, the training process adjusts the connection weights between the connected units to determine to what extent one unit will affect the others. ANNs are successfully employed in NIDSes, which usually fall into two categories: supervised training-based NIDSes and unsupervised training-based NIDSes [69]. As demonstrated in Fig. 6.2, both types of NIDSes essentially follow the architecture and three general steps of ML-NIDS as specified in the beginning of this section.

If the supervised learning approach is applied, the desired output or pattern for a given input is learned from a set of labeled data. A well-known supervised neural network architecture is the multilayer perception (MLP), which is based on the feed-forward and backpropagation algorithms with one or more layers between the input and the output layer [1]. In this type of ANN-NIDS, the number of nodes in the input layer is set to the number of features selected from the original traffic flow, and the number of nodes in the output layer is configured to be the number of desired output classes [16, 70–73]. The number of hidden layers and the number of nodes for each hidden layer vary, and are usually configured according to the situation. A feed-forward-based MLP with a signal hidden layer ANN NIDS model is illustrated in Fig. 6.6.

Obviously, the entire data flow in the ANN is in one direction only: from the input layer, though the hidden layer, to the output layer (see Fig. 6.6). Therefore, given a network traffic package as the input, the corresponding network behavior can be predicted. The advantages of this model are its ability to represent both linear and non-linear relationships, and directly learn these relationships from the data by means of training. However, a number of research projects have reported that the training process of this type of ANN can be very time-consuming, which may pose a significant adverse impact for NIDS system updating [1, 24].

Another group of ANN NIDSes is based on unsupervised training, in which the network adapts to different clusters without having a desired output. One of the most popular algorithms in this group is the self-organizing map (SOM), which transforms the input of arbitrary dimension into a low-dimensional (usually 1- or 2-dimensional) discrete map by using Kohonen's unsupervised learning method [74]. The structure of a conventional self-organizing map is shown in Fig. 6.7a. A conventional SOM network model usually has two layers: an input layer and an output layer (also known as a competitive layer). Similar to the supervised training-based NIDS, the number of nodes in the input layer are usually set to the number of selected features of the

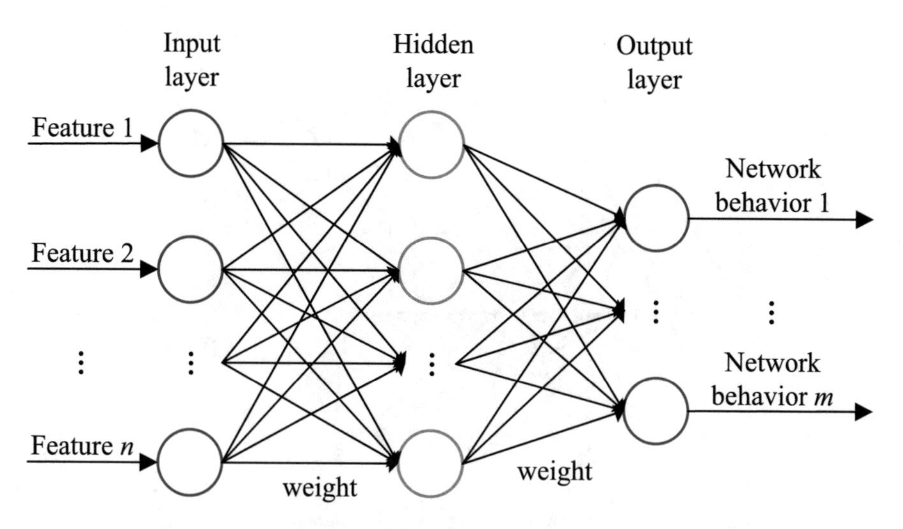

Fig. 6.6 Multilayer perception-based NIDS architecture

training dataset. The output layer consists of neurons organized in a lattice, usually a finite two-dimensional space. Each neuron has a specific topological position and is associated with a weight vector of the same dimension as the input vectors [75].

The training process adjusts the weight vectors of the neurons, thereby describing a mapping from a higher-dimensional input space to a lower-dimensional map space. As a result, the SOM eventually settles into a map of stable zones as a type of feature map of the input space. Based on these mappings, various traffic behaviors can be identified. Figure 6.7b illustrates an example of a SOM output, which clearly shows the four classes that have eventually been predicted.

When comparing the performance (speed and conversion rate) between SOM and supervised learning-based NIDS systems, it becomes clear that SOM is more suitable for real-time intrusion detection [76–80].

Although both types of ANN-network intrusion detection systems are successfully employed in detecting intrusions in real-world network environments with promising results, existing ANN-network intrusion detection systems have two main drawbacks: 1) lower detection precision for low-frequency attacks, and 2) weaker detection stability, which limits the applicability of such systems [16]. The reason behind these is the uneven distribution of different attack types. For example, the number of training data instances for low-frequency attacks are very limited compared to common attacks. As a consequence, it is not easy for the ANN to learn the characteristics of such low-frequency attacks [81].

To address these issues, a number of solutions have been proposed (e.g., [16, 82, 83]). Among these systems, a fuzzy clustering-based neural network NIDS approach (FC-ANN-NIDS) [16] can be a potential solution. Comparing to conventional ANN-NIDSes, in which data clustering techniques are typically not involved during the training process, FC-ANN-NIDSes adopt a fuzzy clustering technique to generate

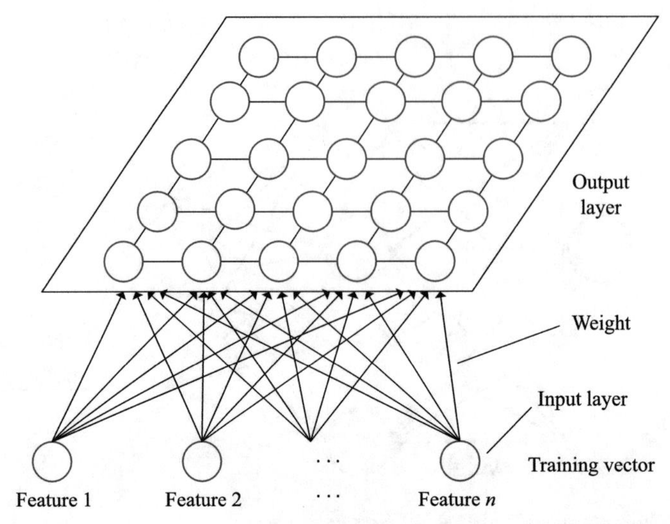

(a) Structure of self-organising map-based NIDSes

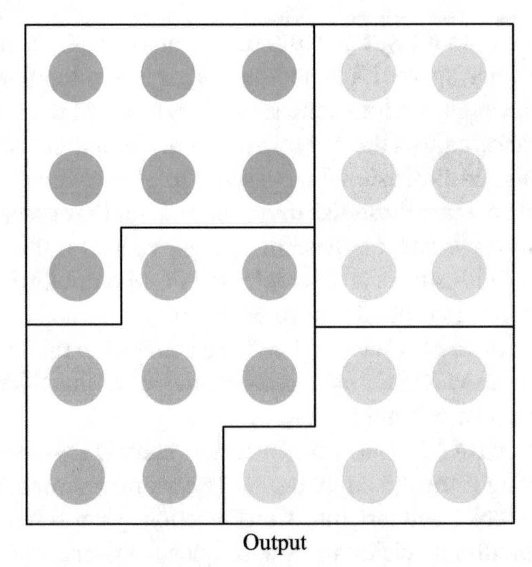

(b) SOM output example

Fig. 6.7 Self-organizing map-based NIDS architecture

different training sub-datasets. This is followed by the application of multiple ANNs in the training stage based on the divided sub-datasets. Finally, a fuzzy aggregation module is applied to combine the results of the ANNs, in an effort to eliminate their errors. The framework of FC-ANN-NIDS is illustrated in Fig. 6.8, which basically

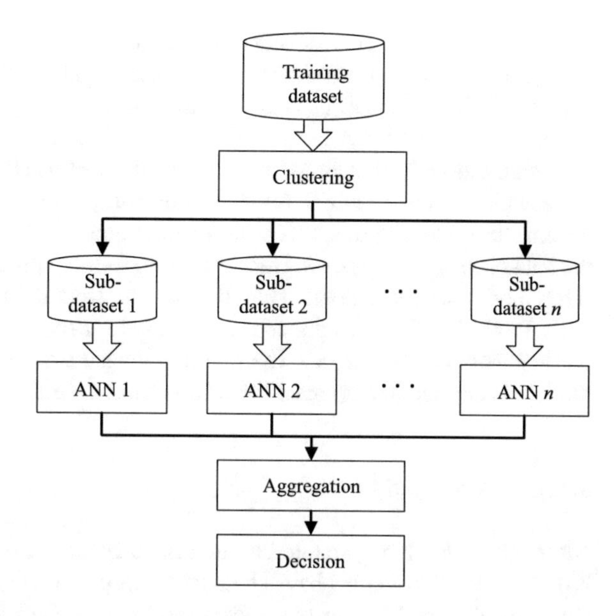

Fig. 6.8 FC-ANN-NIDS framework

contains three major stages: clustering, ANN modeling, and fuzzy aggregation. The details of this method (or FC-ANN-NIDS) are presented in the rest of this section.

6.3.2.1 Clustering

Given a training dataset that contains l network behaviors, the fuzzy C-means clustering technique [84] is employed to group the data instances in clusters, which essentially divides the entire training dataset into n sub-datasets. Note that only the size and complexity of the original training dataset is reduced after data clustering, and the data instances in each divided sub-dataset may still cover all the l network behaviors. Each divided training dataset will be forwarded to the next stage for ANN training. Unfortunately, the value of n (the number of clusters) in the proposed system is determined under a practice theory. Therefore, more intelligent methods, such as the Elbow method [67] may be considered for determining the value of n.

6.3.2.2 ANN Training

A multi-layer perceptron model, illustrated in Fig. 6.6, is used in this study for modeling each sub-training dataset. As mentioned previously, the number of input nodes is set to match the number of selected features of the training dataset; and the number of nodes in the output layer is set to the number of network traffic behaviors covered by the training dataset. The number of hidden nodes is then obtained by adopting the empirical formula: $\sqrt{I+O}+\alpha$, $(\alpha = \{1, \ldots, 10\})$, where I denotes the number of input nodes, O represents the number of nodes in the output layer, and α is a random number [81]. During the training process, the signals, which combine both the input values and the weight values between the corresponding input node and the hidden node, are received by each node in the hidden layer. These signals are processed by a sigmoid activation function, and broadcasted to all the neurons in the output layer with a special weight value. In this study, the most widely used first-order optimization algorithm, gradient descent, is employed for weight-updating during the backpropagation process. Once the entire training process is completed, multiple ANN models can be generated based on the different training sub-datasets. Note that each ANN model can be applied individually for network intrusion detection in real-world network environments. In order to reduce the detection errors, an aggregation module is applied to aggregate the results from different ANNs.

6.3.2.3 Aggregation

Although each ANN generated in the last stage can be deployed individually as an NIDS, some of them may have an unacceptably poor detection performance. In this study, another multi-layer perceptron model is applied for sub-result aggregation. In this stage, the number of nodes in both the input and the output layer is set to the number of network behaviors. Given the entire training dataset and the multiple trained ANN models with the corresponding training sub-datasets generated in the last stage, the modeling process in the aggregation stage can be summarized as follows:

Step 1: Feed each data instance j in the original training dataset to every trained ANN model ($ANN_1, ANN_2, \ldots, ANN_n$). Denote the output of model ANN_i, ($i = \{1, \ldots, n\}$) from data instance j as o_i^j, then the outputs from all ANNs collectively as O^j and $O^j = [o_1^j, \ldots, o_n^j]$.

Step 2: Form the new input for the new ANN model based on the previous outputs. The new input I_{new}^j generated from data instance j is

$$I_{New}^j = [o_1^j \cdot \mu_1, \ldots, o_n^j \cdot \mu_n] \,, \tag{6.7}$$

where μ_i represents the degree of membership of data instance j belonging to cluster i. Note that the degree of membership for each data instance regarding each cluster has been determined in the clustering using the fuzzy C-means clustering algorithm.

Step 3: Generate a new ANN model and train it using the newly formed inputs generated in Step 2.

Once the entire model is built, the system can be deployed in real-world network environments for intrusion detection. Given an incoming network traffic package, the system first calculates the membership of the incoming data using the cluster centers obtained in the first stage. Next, the ANN models and the aggregation model will be applied to predict the final result, which indicates whether the incoming traffic poses a threat. Such hybrid ANN network intrusion detection solutions can increase detection performance, especially for low-frequency attacks. However, they may be costly in time because of the training processes for the large number of feed-forward neural networks.

6.3.3 Deployment of ML-Based NIDSes

Although the developed ML-based network intrusion detection systems are able to take the network package (input) to predict whether it is a normal network behavior, these systems still cannot be directly implemented in real-world network environments for real-time detection. The reason behind this is that the generated ML-based models do not have packet sniffers, which are used to capture the network traffic in real time. In order to achieve real-time detection, the developed ML-based network intrusion detection systems have to work with packet sniffers, such as Snort, Bro, or Spark. A packet sniffer (or network sniffer) is a network traffic monitoring and analyzing tool that can sniff out the network data traversing the monitored network in real time. A number of ML-based network intrusion detection systems have been successfully integrated with packet sniffers and achieved good real-time detection (e.g., [34, 85]). The general framework of these systems is illustrated in Fig. 6.9. A packet sniffer, which can be implemented by either a passive or an in-line deployment method as introduced in Sect. 6.2.1, continuously captures the network traffic, and extracts the required information from the captured network packets to feed into the system model developed by machine learning techniques, thereby generating the final decisions.

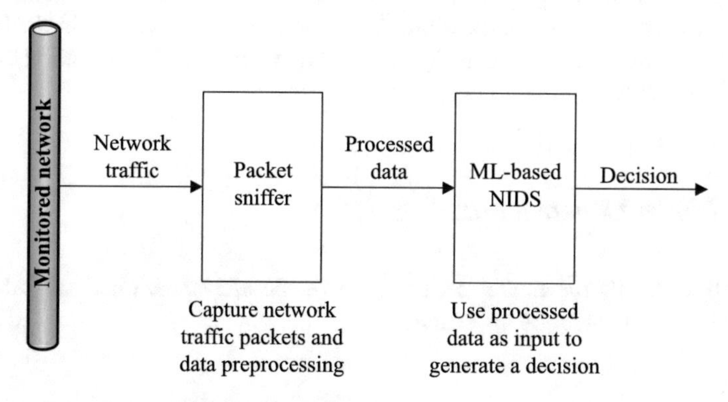

Fig. 6.9 The framework of ML-NIDS deployment

6.4 Experiment

A number of network intrusion detection systems developed by different machine learning approaches are evaluated in this section by applying them to the KDD 99 benchmark dataset.

6.4.1 *Evaluation Environment*

A well-known benchmark dataset, KDD 99, which has been utilized a number of times in recent research [14, 16, 18, 86], is used in this work to evaluate multiple machine learning-based network intrusion detection systems. The KDD 99 dataset is a popular benchmark for intrusion detection; it includes legitimate connections and a wide variety of intrusions simulated in a military network environment [87]. This dataset contains almost 5 million data instances with 42 attributes, including the "class" attribute, which indicates whether a given instance is a normal connection instance or one of the four types of attacks to be identified (i.e., normal, denial of service attacks, user-to-root attacks, remote-to-user attacks, and probes). An important feature of this dataset is that it is an imbalanced dataset, with most data instances belonging to the normal, denial of service attack, and probe categories. As is the case of low-frequency attacks, the classes of user-to-root attacks and remote-to-user attacks are only covered by a small number of data samples. Knowing the inherent issues associated with the dataset, such as the high duplication rate of 78% [87], data instance selection methods, such as the random selection method, are used to reduce the size of the dataset for machine learning. It is worth mentioning that the KDD 99 dataset has been succeeded by the NSL-KDD-99 dataset [87], which reduces the size to 125,937 data samples, while keeping all the features of the original dataset. Table 6.2 details the information about the number of data instances in the training and testing datasets that were used by different network intrusion detection systems, as discussed in Sect. 6.3.

6.4.2 *Model Construction*

This section details the model construction of the aforementioned six ML-based network intrusion detection approaches.

6.4.2.1 TSK+ Fuzzy Inference

As discussed in Sect. 6.3.1, this system brings four important features to the system model. During rule base initialization, the training dataset was divided into five sub-datasets based on the five symbolic labels, which are represented by five integer numbers. The fuzzy model takes four inputs, and predicts a crisp number. According to the Elbow method, 46 TSK fuzzy rules have been generated, which constructed the initial rule base. The final rule base has then been optimized using the GA. The objective function in this work is defined as the root mean square error (RMSE), while the GA parameters are listed in Table 6.3.

6.4.2.2 Conventional Mamdani Fuzzy Inference

The conventional Mamdani fuzzy inference model is investigated in this section. The system uses 34 features for system modeling, which results in 34 inputs and one output Mamdani fuzzy model. Each input domain has been equally partitioned into four regions, described by four linguistic terms, namely, "very low," "low," "medium," and "high;" and two fuzzy sets, "low" and "high," are used to indicate normal and abnormal network traffic, respectively. The fuzzy rules are obtained by a mapping mechanism based on the given training dataset. Given the input, which is a network traffic package, the system first fuzzifies the crisp value of the required features based on the mapping mechanism, then generates a fuzzy output based on

Table 6.2 Details of datasets for machine learning-based NIDSes

Machine learning approach	Training		Testing		Dataset
	Normal	Abnormal	Normal	Abnormal	
TSK+ [14]	67,343	58,630	9,711	9,083	Entire NSL-KDD-99
Conventional fuzzy inference [33]	67,343	58,630	9,711	9,083	Entire NSL-KDD-99
FC-ANN [16]	3,000	15,285	60,593	250,496	Random part
MLP [24]	5,922	6,237	3,608	3,388	Random part
SOM [10]	97,277	396,744	60,593	250,436	Random part
Hierarchical SOM [10]	97,277	396,744	60,593	250,436	Random part

Table 6.3 GA parameters

Parameters	Values
Population size	100.00
Crossover rate	0.85
Mutation rate	0.05
Maximum iteration	10,000.00
Termination threshold	0.01

the generated rule base. Finally, the center of gravity method is employed to defuzzify the fuzzy output to a crisp one, which indicates whether the traffic is normal or attack traffic.

6.4.2.3 Fuzzy Clustering-Based ANN

Fuzzy clustering-based ANN uses all the 41 features to predict the five network behaviors. Note that the symbolic values contained in the dataset have been converted to continuous values. In the beginning, six training sub-datasets are obtained by using fuzzy C-means clustering. From there, six signal-hidden-layered neural network models are trained, each of which is referred to as the [41;18;5] structure. This means that each network takes 41 inputs, goes through 18 hidden nodes, and finally produces 5 outputs. In the aggregation progress, a new signal hidden-layered ANN model with the structure [5;13;5] is designed to aggregate all results from upper-level ANN models. The mean square error (MSE) is used as the fitness function during system modeling, and the threshold of MSE is set to 0.001. Also, the learning rate and the momentum factor at both ANN model levels are set to 0.01 and 0.2, respectively.

6.4.2.4 Multilayer Perceptron

Expert knowledge has been used in this work to help select the most important features. In particular, 35 features, including five symbolic features and 30 numerical features, have been selected. Similar to the FC-ANN approach introduced above, the symbolic values were converted to numerical values. Because of the lack of data samples in U2R and R2U attacks, only three categories, namely, "normal," "DoS," and "probes," were considered. As a result, 35 input nodes and three output nodes were used. In this experiment, a two hidden-layered MLP network model was implemented, constituting a four-layer MLP, whose structure is referred to as [35;35;35;3].

6.4.2.5 Hierarchical Self-Organizing Maps

A hierarchical self-organizing map architecture, which consists of two levels of SOM networks, each comprised of three layers, was used in this experiment. The first layer was an input layer, with 20 input nodes (corresponding to 20 selected features). At the first level of SOM, six SOM networks were deployed, each of which represented one of the basic TCP features, including "duration," "protocol type," "service," "flag," "destination bytes," and "source bytes." During the training process, each training data sample was fed into each SOM network, thereby creating a number of mappings between inputs and six 6×6 grids on the second layer, which resulted in $36 \times 6 = 216$ neurons. After this, *potential function clustering* [84] was employed on each output layer of the first SOM level to reduce the total neurons from 36 to 6. As a consequence, the total number of neurons in the second layer was reduced to 36. These 36 neurons were used as inputs for the second SOM level to train a new SOM network that consists of a 20×20 grid of neurons, which indicates the mapping from the input space to the different network behaviors. The learning rate was set to 0.05, and the neighborhood function was configured as a Gaussian function.

6.4.2.6 Conventional Self-Organizing Maps

In this experiment, all the 41 features have been used for the intrusion detection system. During the training process, the learning rate was set to 0.05, and the Gaussian function was used as the neighborhood function. The developed system took 41 inputs to create a mapping between five categories of network behaviors into a 6×6 grid of neurons.

6.4.3 Result Comparisons

In order to enable a direct comparison between the different ML-NIDS approaches, a common measurement, the detection rate, is employed in this work. In particular, the detection rate can be defined as follows:

$$\text{Detection rate} = \frac{\text{Number of instances correctly detected}}{\text{Total number of instances}} \cdot 100 \tag{6.8}$$

The detection rates of the classification results for each network traffic category are summarized in Table 6.4.

The results show that all the approaches achieved a high detection performance in the normal, DoS, and probes category, which contain sufficient data samples for training. Note that conventional ANN-based network intrusion detection systems, such as the MLP-based approach and the SOM-based approach, led to an extremely poor detection performance in the case of U2R and R2U. As discussed in Sect. 6.3.2,

Table 6.4 Performance comparison

Approach	Normal	DoS	U2R	R2U	Probes
TSK+ [14]	93.10	97.84	65.38	84.65	85.69
Conventional fuzzy inference [33]	82.93	90.42	19.05	15.58	37.08
FC-ANN [16]	91.32	96.70	76.92	58.57	80.00
MLP [24]	89.20	90.90	N/A	N/A	90.30
SOM [10]	98.50	96.80	0.00	0.15	63.40
Hierarchical SOM [10]	92.40	96.50	22.90	11.30	72.80

this issue is caused by the lack of training data samples for both U2R and R2U. In this case, a future investigation may be required to identify how the detection threshold affects the detection performance. Obviously, similar to the modified version of the ANN approaches, the FC-ANN-based approach and the hierarchical SOM-based approach increased the detection rate. It is worth mentioning that the TSK− based intrusion detection system not only achieved the best detection performance in the normal, DoS, and probes classes, but also had an outstanding performance in the other two classes.

6.5 Conclusion

This chapter investigated how machine learning algorithms can be used to develop NIDSes. In particular, the chapter first reviewed the existing intrusion detection techniques, including hardware deployments and software implementations. They are followed by the discussion of a number of machine learning algorithms and their applications in network intrusion detection. Finally, a well-known network security benchmark dataset, KDD 99, was employed for the evaluation of the reviewed machine learning-based network intrusion detection systems, with a critical analysis of the results. Although the benchmark dataset, KDD 99, is still popular in recent research, it is relatively outdated and many of today's network threats are not covered by the KDD 99 dataset. Therefore, future research may consider using alternate datasets (e.g., [88, 89]). In addition, as IoT continues to expand, the data being generated will continue to grow in volume and velocity. How conventional machine learning and artificial intelligence techniques can be expanded to deal with the continuously growing data is an interesting research direction.

References

1. Stampar M, Fertalj K (2015) Artificial intelligence in network intrusion detection. In: Biljanovic P, Butkovic Z, Skala K, Mikac B, Cicin-Sain M, Sruk V, Ribaric S, Gros S, Vrdoljak B, Mauher M, Sokolic A (eds) Proceedings of the 38th International Convention on Information and Communication Technology, Electronics and Microelectronics, pp 1318–1323. https://doi.org/10.1109/MIPRO.2015.7160479
2. Sommer R, Paxson V (2010) Outside the closed world: on using machine learning for network intrusion detection. In: Proceedings of the 2010 IEEE Symposium on Security and Privacy. IEEE Computer Society, Los Alamitos, CA, USA, pp 305–316. https://doi.org/10.1109/SP.2010.25
3. Buczak AL, Guven E (2016) A survey of data mining and machine learning methods for cyber security intrusion detection. IEEE Commun Surv Tutor 18(2):1153–1176. https://doi.org/10.1109/COMST.2015.2494502
4. Russell SJ, Norvig P (2009) Artificial intelligence: a modern approach, 3rd edn. Pearson, Essex
5. Farnaaz N, Jabbar M (2016) Random forest modeling for network intrusion detection system. Procedia Comput Sci 89:213–217. https://doi.org/10.1016/j.procs.2016.06.047

6. Ma Z, Kaban A (2013) K-nearest-neighbours with a novel similarity measure for intrusion detection. In: Jin Y, Thomas SA (eds) Proceedings of the 13th UK Workshop on Computational Intelligence. IEEE, New York, pp 266–271. https://doi.org/10.1109/UKCI.2013.6651315
7. Mukherjee S, Sharma N (2012) Intrusion detection using Naïve Bayes classifier with feature reduction. Proc Tech 4:119–128. https://doi.org/10.1016/j.protcy.2012.05.017
8. Thaseen IS, Kumar CA (2017) Intrusion detection model using fusion of chi-square feature selection and multi class SVM. J King Saud Univ Comput Inf Sci 29(4):462–472. https://doi.org/10.1016/j.jksuci.2015.12.004
9. Zhang C, Zhang G, Sun S (2009) A mixed unsupervised clustering-based intrusion detection model. In: Huang T, Li L, Zhao M (eds) Proceedings of the Third International Conference on Genetic and Evolutionary Computing. IEEE Computer Society, Los Alamitos, CA, USA, pp 426–428. https://doi.org/10.1109/WGEC.2009.72
10. Kayacik HG, Zincir-Heywood AN, Heywood MI (2007) A hierarchical SOM-based intrusion detection system. Eng Appl Artif Intell 20(4):439–451. https://doi.org/10.1016/j.engappai.2006.09.005
11. Garfinkel S (2002) Network forensics: tapping the Internet. https://paulohm.com/classes/cc06/files/Week6%20Network%20Forensics.pdf
12. Liao HJ, Lin CHR, Lin YC, Tung KY (2013) Intrusion detection system: a comprehensive review. J Netw Comput Appl 36(1):16–24. https://doi.org/10.1016/j.jnca.2012.09.004
13. Bostani H, Sheikhan M (2017) Modification of supervised OPF-based intrusion detection systems using unsupervised learning and social network concept. Pattern Recogn 62:56–72. https://doi.org/10.1016/j.patcog.2016.08.027
14. Li J, Yang L, Qu Y, Sexton G (2018) An extended Takagi-Sugeno-Kang inference system (TSK+) with fuzzy interpolation and its rule base generation. Soft Comput 22(10):3155–3170. https://doi.org/10.1007/s00500-017-2925-8
15. Ramadas M, Ostermann S, Tjaden B (2003) Detecting anomalous network traffic with self-organizing maps. In: Vigna G, Krügel C, Jonsson E (eds) Recent advances in intrusion detection. Springer, Heidelberg, pp 36–54. https://doi.org/10.1007/978-3-540-45248-5_3
16. Wang G, Hao J, Ma J, Huang L (2010) A new approach to intrusion detection using artificial neural networks and fuzzy clustering. Expert Syst Appl 37(9):6225–6232. https://doi.org/10.1016/j.eswa.2010.02.102
17. Wang W, Battiti R (2006) Identifying intrusions in computer networks with principal component analysis. In: Revell N, Wagner R, Pernul G, Takizawa M, Quirchmayr G, Tjoa AM (eds) Proceedings of the First International Conference on Availability, Reliability and Security. IEEE Computer Society, Los Alamitos, CA, USA. https://doi.org/10.1109/ARES.2006.73
18. Yang L, Li J, Fehringer G, Barraclough P, Sexton G, Cao Y (2017) Intrusion detection system by fuzzy interpolation. In: Proceedings of the 2017 IEEE International Conference on Fuzzy Systems. https://doi.org/10.1109/FUZZ-IEEE.2017.8015710
19. Sekar R, Gupta A, Frullo J, Shanbhag T, Tiwari A, Yang H, Zhou S (2002) Specification-based anomaly detection: a new approach for detecting network intrusions. In: Proceedings of the 9th ACM Conference on Computer and Communications Security. ACM, New York, pp 265–274. https://doi.org/10.1145/586110.586146
20. Tseng CY, Balasubramanyam P, Ko C, Limprasittiporn R, Rowe J, Levitt K (2003) A specification-based intrusion detection system for AODV. In: Swarup V, Setia S (eds) Proceedings of the 1st ACM Workshop on Security of ad hoc and Sensor Networks. ACM, New York, pp 125–134. https://doi.org/10.1145/986858.986876
21. Bostani H, Sheikhan M (2017) Hybrid of anomaly-based and specification-based IDS for Internet of Things using unsupervised OPF based on MapReduce approach. Comput Commun 98:52–71. https://doi.org/10.1016/j.comcom.2016.12.001
22. Mukkamala S, Sung A (2003) Feature selection for intrusion detection with neural networks and support vector machines. Trans Res Rec 1822:33–39. https://doi.org/10.3141/1822-05
23. Kumar M, Hanumanthappa M, Kumar TVS (2012) Intrusion detection system using decision tree algorithm. In: Proceedings of the 14th IEEE International Conference on Communication Technology. IEEE, New York, pp 629–634. https://doi.org/10.1109/ICCT.2012.6511281

24. Moradi M, Zulkernine M (2004) A neural network based system for intrusion detection and classification of attacks. http://research.cs.queensu.ca/~moradi/148-04-MM-MZ.pdf
25. Ravale U, Marathe N, Padiya P (2015) Feature selection based hybrid anomaly intrusion detection system using K means and RBF kernel function. Procedia Comput Sci 45:428–435. https://doi.org/10.1016/j.procs.2015.03.174
26. Liu G, Yi Z (2006) Intrusion detection using PCASOM neural networks. In: Wang J, Yi Z, Zurada JM, Lu BL, Yin H (eds) Advances in neural networks–ISNN 2006. Springer, Heidelberg, pp 240–245. https://doi.org/10.1007/11760191_35
27. Chen Y, Abraham A, Yang B (2007) Hybrid flexible neural-tree-based intrusion detection systems. Int J Intell Syst 22(4):337–352. https://doi.org/10.1002/int.20203
28. Mamdani EH (1977) Application of fuzzy logic to approximate reasoning using linguistic synthesis. IEEE Trans Comput C-26(12):1182–1191. https://doi.org/10.1109/TC.1977.1674779
29. Takagi T, Sugeno M (1985) Fuzzy identification of systems and its applications to modeling and control. IEEE Trans Syst Man Cybern SMC-15(1):116–132. https://doi.org/10.1109/TSMC.1985.6313399
30. Li J, Shum HP, Fu X, Sexton G, Yang L (2016) Experience-based rule base generation and adaptation for fuzzy interpolation. In: Cordón O (ed) Proceedings of the 2016 IEEE International Conference on Fuzzy Systems. IEEE, New York, pp 102–109. https://doi.org/10.1109/FUZZ-IEEE.2016.7737674
31. Tan Y, Li J, Wonders M, Chao F, Shum HP, Yang L (2016) Towards sparse rule base generation for fuzzy rule interpolation. In: Cordón O (ed) Proceedings of the 2016 IEEE International Conference on Fuzzy Systems. IEEE, New York, pp 110–117. https://doi.org/10.1109/FUZZ-IEEE.2016.7737675
32. Chaudhary A, Tiwari V, Kumar A (2014) Design an anomaly based fuzzy intrusion detection system for packet dropping attack in mobile ad hoc networks. In: Batra U (ed) Proceedings of the 2014 IEEE International Advance Computing Conference. IEEE, New York, pp 256–261. https://doi.org/10.1109/IAdCC.2014.6779330
33. Shanmugavadivu R, Nagarajan N (2011) Network intrusion detection system using fuzzy logic. Indian J Comput Sci Eng 2(1):101–111
34. Naik N, Diao R, Shen Q (2017) Dynamic fuzzy rule interpolation and its application to intrusion detection. IEEE Trans Fuzzy Syst https://doi.org/10.1109/TFUZZ.2017.2755000
35. Kóczy TL, Hirota K (1993) Approximate reasoning by linear rule interpolation and general approximation. Int J Approx Reason 9(3):197–225. https://doi.org/10.1016/0888-613X(93)90010-B
36. Huang Z, Shen Q (2006) Fuzzy interpolative reasoning via scale and move transformations. IEEE Trans Fuzzy Syst 14(2):340–359. https://doi.org/10.1109/TFUZZ.2005.859324
37. Huang Z, Shen Q (2008) Fuzzy interpolation and extrapolation: a practical approach. IEEE Trans Fuzzy Syst 16(1):13–28. https://doi.org/10.1109/TFUZZ.2007.902038
38. Li J, Yang L, Fu X, Chao F, Qu Y (2018) Interval Type-2 TSK+ fuzzy inference system. In: Proceedings of the 2018 IEEE International Conference on Fuzzy Systems. Curran Associates, Red Hook, NY, USA
39. Yang L, Shen Q (2010) Adaptive fuzzy interpolation and extrapolation with multiple-antecedent rules. In: Proceedings of the 2010 IEEE International Conference on Fuzzy Systems. Curran Associates, Red Hook, NY, USA. https://doi.org/10.1109/FUZZY.2010.5584701
40. Naik N, Diao R, Quek C, Shen Q (2013) Towards dynamic fuzzy rule interpolation. In: Proceedings of the 2013 IEEE International Conference on Fuzzy Systems. Curran Associates, Red Hook, NY, USA. https://doi.org/10.1109/FUZZ-IEEE.2013.6622404
41. Naik N, Diao R, Shen Q (2014) Genetic algorithm-aided dynamic fuzzy rule interpolation. In: Proceedings of the 2014 IEEE International Conference on Fuzzy Systems. Curran Associates, Red Hook, NY, USA. https://doi.org/10.1109/FUZZ-IEEE.2014.6891816
42. Shen Q, Yang L (2011) Generalisation of scale and move transformation-based fuzzy interpolation. J Adv Comput Intell Int Inf 15(3):288–298. https://doi.org/10.20965/jaciii.2011.p0288
43. Yang L, Chao F, Shen Q (2017) Generalised adaptive fuzzy rule interpolation. IEEE Trans Fuzzy Syst 25(4):839–853. https://doi.org/10.1109/TFUZZ.2016.2582526

44. Yang L, Chen C, Jin N, Fu X, Shen Q (2014) Closed form fuzzy interpolation with interval type-2 fuzzy sets. In: Proceedings of the 2014 IEEE International Conference on Fuzzy Systems. IEEE, pp 2184–2191. https://doi.org/10.1109/FUZZ-IEEE.2014.6891643
45. Yang L, Shen Q (2011) Adaptive fuzzy interpolation. IEEE Trans Fuzzy Syst 19(6):1107–1126. https://doi.org/10.1109/TFUZZ.2011.2161584
46. Yang L, Shen Q (2011) Adaptive fuzzy interpolation with uncertain observations and rule base. In: Lin C-T, Kuo Y-H (eds) Proceedings of the 2011 IEEE International Conference on Fuzzy Systems. IEEE, New York, pp 471–478. https://doi.org/10.1109/FUZZY.2011.6007582
47. Yang L, Shen Q (2013) Closed form fuzzy interpolation. Fuzzy Sets Syst 225:1–22. https://doi.org/10.1016/j.fss.2013.04.001
48. Li J, Yang L, Fu X, Chao F, Qu Y (2017) Dynamic QoS solution for enterprise networks using TSK fuzzy interpolation. In: Proceedings of the 2017 IEEE International Conference on Fuzzy Systems. Curran Associates, Red Hook, NY, USA. https://doi.org/10.1109/FUZZ-IEEE.2017.8015711
49. Li J, Yang L, Shum HP, Sexton G, Tan Y (2015) Intelligent home heating controller using fuzzy rule interpolation. In: UK Workshop on Computational Intelligence, 7–9 September 2015, Exeter, UK
50. Naik N (2015) Fuzzy inference based intrusion detection system: FI-Snort. In: Wu Y, Min G, Georgalas N, Hu J, Atzori L, Jin X, Jarvis S, Liu L, Calvo RA (eds) Proceedings of the 2015 IEEE International Conference on Computer and Information Technology; Ubiquitous Computing and Communications; Dependable, Autonomic and Secure Computing; Pervasive Intelligence and Computing. IEEE Computer Society, Los Alamitos, CA, USA, pp 2062–2067. https://doi.org/10.1109/CIT/IUCC/DASC/PICOM.2015.306
51. Yang L, Li J, Hackney P, Chao F, Flanagan M (2017) Manual task completion time estimation for job shop scheduling using a fuzzy inference system. In: Wu Y, Min G, Georgalas N, Al-Dubi A, Jin X, Yang L, Ma J, Yang P (eds) Proceedings of the 2017 IEEE International Conference on Internet of Things (iThings) and IEEE Green Computing and Communications (GreenCom) and IEEE Cyber, Physical and Social Computing (CPSCom) and IEEE Smart Data (SmartData). IEEE Computer Society, Los Alamitos, CA, USA, pp 139–146. https://doi.org/10.1109/iThings-GreenCom-CPSCom-SmartData.2017.26
52. Li J, Qu Y, Shum HPH, Yang L (2017) TSK inference with sparse rule bases. In: Angelov P, Gegov A, Jayne C, Shen Q (eds) Advances in computational intelligence systems. Springer, Cham, pp 107–123. https://doi.org/10.1007/978-3-319-46562-3_8
53. Guha S, Yau SS, Buduru AB (2016) Attack detection in cloud infrastructures using artificial neural network with genetic feature selection. In: Proceedings of the 14th International Conference on Dependable, Autonomic and Secure Computing, 14th International Conference on Pervasive Intelligence and Computing, 2nd International Conference on Big Data Intelligence and Computing and Cyber Science and Technology Congress. IEEE Computer Society, Los Alamitos, CA, USA, pp 414–419. https://doi.org/10.1109/DASC-PICom-DataCom-CyberSciTec.2016.32
54. Jensen R, Shen Q (2008) Computational intelligence and feature selection: rough and fuzzy approaches. Wiley-IEEE Press, New York
55. Jensen R, Shen Q (2009) New approaches to fuzzy-rough feature selection. IEEE Trans Fuzzy Syst 17(4):824–838. https://doi.org/10.1109/TFUZZ.2008.924209
56. Tsang EC, Chen D, Yeung DS, Wang XZ, Lee JW (2008) Attributes reduction using fuzzy rough sets. IEEE Trans Fuzzy Syst 16(5):1130–1141. https://doi.org/10.1109/TFUZZ.2006.889960
57. Zuo Z, Li J, Anderson P, Yang L, Naik N (2018) Grooming detection using fuzzy-rough feature selection and text classification. In: Proceedings of the 2018 IEEE International Conference on Fuzzy Systems. Curran Associates, Red Hook, NY, USA
58. Dash M, Liu H (1997) Feature selection for classification. Intell. Data Anal 1(3):131–156. https://doi.org/10.1016/S1088-467X(97)00008-5
59. Langley P (1994) Selection of relevant features in machine learning. In: Proceedings of the AAAI Fall Symposium on Relevance. AAAI Press, Palo Alto, CA, USA, pp 245–271

60. Jensen R, Shen Q (2009) Are more features better? A response to attributes reduction using fuzzy rough sets. IEEE Trans Fuzzy Syst 17(6):1456–1458. https://doi.org/10.1109/TFUZZ.2009.2026639
61. Guyon I, Elisseeff A (2003) An introduction to variable and feature selection. J Mach Learn Res 3:1157–1182. http://www.jmlr.org/papers/volume3/guyon03a/guyon03a.pdf
62. Jensen R, Shen Q (2004) Semantics-preserving dimensionality reduction: rough and fuzzy-rough-based approaches. IEEE Trans Knowl Data Eng 16(12):1457–1471. https://doi.org/10.1109/TKDE.2004.96
63. Parthaláin NM, Shen Q (2009) Exploring the boundary region of tolerance rough sets for feature selection. Pattern Recogn 42(5):655–667. https://doi.org/10.1016/j.patcog.2008.08.029
64. Parthaláin NM, Shen Q, Jensen R (2010) A distance measure approach to exploring the rough set boundary region for attribute reduction. IEEE Trans Knowl Data Eng 22(3):305–317. https://doi.org/10.1109/TKDE.2009.119
65. Saeys Y, Inza I, Larrañaga P (2007) A review of feature selection techniques in bioinformatics. Bioinformatics 23(19):2507–2517. https://doi.org/10.1093/bioinformatics/btm344
66. Yu L, Liu H (2004) Efficient feature selection via analysis of relevance and redundancy. J Mach Learn Res 5:1205–1224
67. Thorndike RL (1953) Who belongs in the family? Psychometrika 18(4):267–276. https://doi.org/10.1007/BF02289263
68. Anderson JA (1995) An introduction to neural networks. MIT Press, Cambridge, MA, USA
69. Planquart J-P (2001) Application of neural networks to intrusion detection. Sans Institute. https://www.sans.org/reading-room/whitepapers/detection/application-neural-networks-intrusion-detection-336
70. Cameron R, Zuo Z, Sexton G, Yang L (2017) A fall detection/recognition system and an empirical study of gradient-based feature extraction approaches. In: Chao F, Schockaert S, Zhang Q (eds) Advances in computational intelligence systems. Springer, Cham, pp 276–289. https://doi.org/10.1007/978-3-319-66939-7_24
71. Linda O, Vollmer T, Manic M (2009) Neural network based intrusion detection system for critical infrastructures. In: Proceedings of the 2009 International Joint Conference on Neural Networks. IEEE, Piscataway, NJ, USA, pp 1827–1834. https://doi.org/10.1109/IJCNN.2009.5178592
72. Subba B, Biswas S, Karmakar S (2016) A neural network based system for intrusion detection and attack classification. In: Proceedings of the Twenty-Second National Conference on Communication. IEEE, New York. https://doi.org/10.1109/NCC.2016.7561088
73. Zuo Z, Yang L, Peng Y, Chao F, Qu Y (2018) Gaze-informed egocentric action recognition for memory aid systems. IEEE Access 6:12894–12904. https://doi.org/10.1109/ACCESS.2018.2808486
74. Beghdad R (2008) Critical study of neural networks in detecting intrusions. Comput Secur 27(5):168–175. https://doi.org/10.1016/j.cose.2008.06.001
75. Ouadfel S, Batouche M (2007) Antclust: an ant algorithm for swarm-based image clustering. Inf Technol J 6(2):196–201. https://doi.org/10.3923/itj.2007.196.201
76. De la Hoz E, de la Hoz E, Ortiz A, Ortega J, Martínez-Álvarez A: Feature selection by multi-objective optimisation: application to network anomaly detection by hierarchical self-organising maps. Knowl Based Syst 71:322–338. https://doi.org/10.1016/j.knosys.2014.08.013
77. Labib K, Vemuri R (2002) NSOM: a real-time network-based intrusion detection system using self-organizing maps. http://web.cs.ucdavis.edu/~vemuri/papers/som-ids.pdf
78. Vasighi M, Amini H (2017) A directed batch growing approach to enhance the topology preservation of self-organizing map. Appl Soft Comput 55:424–435. https://doi.org/10.1016/j.asoc.2017.02.015
79. Vokorokos L, Balaz A, Chovanec M (2006) Intrusion detection system using self organizing map. Acta Electrotechnica et Informatica 6(1). http://www.aei.tuke.sk/papers/2006/1/Vokorokos.pdf

80. Prabhakar SY, Parganiha P, Viswanatham VM, Nirmala M (2017) Comparison between genetic algorithm and self organizing map to detect botnet network traffic. In: IOP conference series: materials science and engineering, vol 263. IOP Publishing, Bristol. https://doi.org/10.1088/1757-899X/263/4/042103
81. Haykin S (2009) Neural networks and learning machines, 3rd edn. Prentice Hall, Upper Saddle River, NJ, USA
82. Joo D, Hong T, Han I (2003) The neural network models for IDS based on the asymmetric costs of false negative errors and false positive errors. Expert Syst Appl 25(1):69–75. https://doi.org/10.1016/S0957-4174(03)00007-1
83. Patcha A, Park JM (2007) An overview of anomaly detection techniques: existing solutions and latest technological trends. Comput Netw 51(12):3448–3470. https://doi.org/10.1016/j.comnet.2007.02.001
84. Chiu SL (1994) Fuzzy model identification based on cluster estimation. J Intell Fuzzy Syst 2(3):267–278. https://doi.org/10.3233/IFS-1994-2306
85. Mahoney MV (2003) A machine learning approach to detecting attacks by identifying anomalies in network traffic. Ph.D. thesis, Florida Institute of Technology, Melbourne, FL, USA
86. Elisa N, Yang L, Naik N (2018) Dendritic cell algorithm with optimised parameters using genetic algorithm. In: Proceedings of the 2018 IEEE Congress on Evolutionary Computation. Curran Associates, Red Hook, NY, USA
87. Tavallaee M, Bagheri E, Lu W, Ghorbani A (2009) A detailed analysis of the KDD Cup 99 data set. In: Wesolkowski S, Abbass H, Abielmona R (eds) Proceedings of the 2009 IEEE Symposium on Computational Intelligence for Security and Defense Applications. https://doi.org/10.1109/CISDA.2009.5356528
88. Gharib A, Sharafaldin I, Lashkari AH, Ghorbani AA (2016) An evaluation framework for intrusion detection dataset. In: Joukov N, Kim H (eds) Proceedings of the 2016 International Conference on Information Science and Security. Curran Associates, Red Hook, NY, USA. https://doi.org/10.1109/ICISSEC.2016.7885840
89. Sharafaldin I, Lashkari AH, Ghorbani AA (2018) Toward generating a new intrusion detection dataset and intrusion traffic characterization. In: Mori P, Furnell S, Camp O (eds) Proceedings of the 4th International Conference on Information Systems Security and Privacy, vol 1, pp 108–116. https://doi.org/10.5220/0006639801080116

Chapter 7
Android Application Analysis Using Machine Learning Techniques

Takeshi Takahashi and Tao Ban

Abstract The amount of malware that target Android terminals is growing. Malware applications are distributed to Android terminals in the form of Android Packages (APKs), similar to other Android applications. Analyzing APKs may thus help identify malware. In this chapter, we describe how machine learning techniques can be used to identify Android malware. We begin by looking at the structure of an APK file and introduce techniques for identifying malware. We then describe how data can be collected and analyzed and then used to prepare a dataset. This is done by not only using permission requests and API calls, but also by using application clusters and descriptions as the source. To demonstrate the effectiveness of machine learning techniques for analyzing Android applications, we analyze the performance of support vector machine classification on our dataset and compare it to that of a scheme that does not utilize machine learning. We also evaluate the effectiveness of the features used and further improve the classification performance by removing irrelevant features. Finally, we address several issues and limitations on the use of machine learning techniques for analyzing Android applications.

7.1 Introduction

Android[1] is one of the most widely used operating systems (OSes) for smartphones, with a global market share of 87.7% at the time of writing in terms of sales to end users [1]. Its open specification facilitates the development of applications and their release on the Android application market. However, this makes it difficult to manage Android OSes and applications in a centralized manner. The difficulty of management enables Android malware to be distributed without being discovered.

The amount of Android malware is increasing, and various types of threats exist on the Android platform [2]. For example, Simplocker [3] and LockerPin [4] are two

[1] https://www.android.com

T. Takahashi (✉) · T. Ban
National Institute of Information and Communications Technology, Tokyo, Japan
e-mail: takeshi_takahashi@nict.go.jp

© Springer Nature Switzerland AG 2019

L. F. Sikos (ed.), *AI in Cybersecurity*, Intelligent Systems Reference Library 151,
https://doi.org/10.1007/978-3-319-98842-9_7

ransomware applications for Android. Simplocker encrypts user files while LockerPin changes a device's personal identification number (PIN)[2] lock. The new PIN is not known by the legitimate user, nor by the attacker, so the user will be unable to obtain the PIN even if the requested ransom is paid.

The impact of Android malware is not limited to smartphones. Although Android is currently used mainly on smartphone devices, it will be widely used by Internet of Things (IoT) devices as well. In fact, an Android-based OS for IoT, called "Android Things," which was rebranded as "Brillo," is already available.[3] Consequently, Android malware will progressively affect more than just smartphones.

In this chapter, we introduce techniques for detecting Android malware and describe how machine learning techniques, in particular support vector machines (SVMs) [5, 6], can be used for analyzing Android applications.[4] We also address dataset generation, which is essential for machine learning techniques, because their performance largely depends on the size and quality of the dataset. We use not only permission requests and API calls, but also application categories and descriptions as the data source. Using the generated dataset, we demonstrate the effectiveness of using an SVM for analyzing Android applications by measuring the classification performance of an SVM and comparing it to that of a scheme that does not utilize machine learning. To improve the performance of the SVM further, we evaluate the effectiveness of the features used for analyzing Android applications and remove the non-contributing encoded features from the dataset. We then describe an experiment in which 94.15% classification accuracy was achieved. We close with a discussion of several issues and limitations on the practical use of machine learning techniques in this field.

The remainder of this chapter is structured as follows. Section 7.2 overviews the Android Application Package (APK) to provide the preliminary knowledge necessary for analyzing Android applications. Section 7.3 shows various techniques for identifying malware, including machine learning-based techniques. Section 7.4 describes the dataset used. Section 7.5 introduces a technique for detecting malware using an SVM and evaluates its effectiveness on our dataset. Section 7.6 provides a performance comparison with a non-machine learning-based scheme. Section 7.7 introduces a technique for improving the generalization ability of machine learning techniques by evaluating features and removing irrelevant features from the dataset. Section 7.8 discusses several issues and limitations on the use of machine learning techniques for analyzing Android applications. Finally, Sect. 7.9 concludes with a summary of the key points and potential lines of future research.

[2]A PIN is a numeric or alpha-numeric password or code used for user authentication.

[3]https://developer.android.com/things/

[4]This article is an extended version of work published in [7, 8].

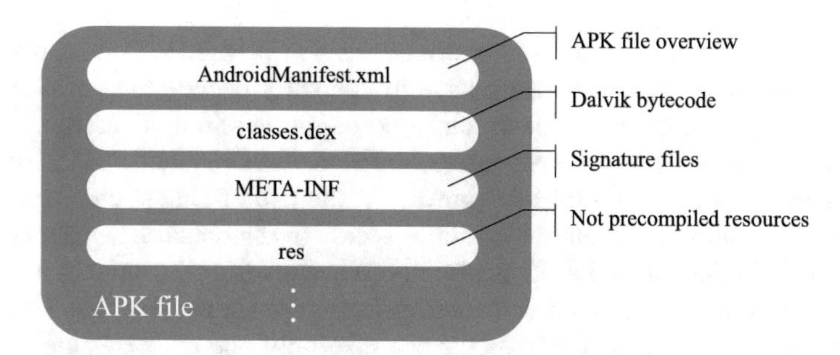

Fig. 7.1 APK file structure

7.2 The Structure of Android Application Packages

Android applications are provided in the form of APKs. An APK file is a ZIP file consisting of multiple files, making it necessary to unzip it before use. As shown in Fig. 7.1, each APK file contains the files `AndroidManifest.xml` and `classes.dex` as well as signatures files and resources that are not precompiled. `AndroidManifest.xml` and `classes.dex` are often used to analyze and evaluate the threats and vulnerabilities of APK files.

7.2.1 Central Configuration (AndroidManifest.xml)

All APK files contain an `AndroidManifest.xml` file, containing assorted application information described in XML, although stored in a binary form. The information can be extracted from this file using tools such as Apktool[5] and Android Studio.[6] Table 7.1 shows some of the tags included in the XML, including, among other things, permission requests and the API levels supported by the application.

Table 7.1 Core information described in AndroidManifest.xml

Tag name	Content
Application	General configuration of application, such as icons, labels, and display theme
Uses-sdk	Range of API levels needed to run the application
Uses-permission	Permissions requested by the application
Uses-library	Libraries used by the application

[5]https://ibotpeaches.github.io/Apktool/

[6]https://developer.android.com/studio/

A permission system is used to restrict access to privileged system resources, and Android application developers have to explicitly declare the permissions in `AndroidManifest.xml`. The official Android permissions are categorized into four types: Normal, Dangerous, Signature, and SignatureOrSystem, the last of which was discontinued from Android 6.0 onwards. The use of Dangerous permissions requires user approval, because they allow access to restricted resources and may have security implications if used incorrectly. When taken as the input of a machine learning algorithm for malware detection, permissions are usually coded as binary variables, i.e., an element in the vector can take only one of two values: 1 for a requested permission and 0 otherwise. The number of possible Android permissions depends on the version of the OS.

The Android OS is continuously evolving, and the available permission requests and API calls may change from time to time. Therefore, one needs to refer to the `uses-sdk` tag to obtain the supported API versions, and run risk analysis using the supported permissions and API calls to conduct risk analysis efficiently.

7.2.2 Dalvik Bytecode (classes.dex)

Android applications are developed in Java and compiled into Java bytecode. The bytecode is then translated into Dalvik bytecode and stored in the Dalvik Executable (DEX) format,[7] i.e., `classes.dex`. Dalvik bytecode is, like Java bytecode, reverse-engineering-friendly, enabling code analysis without the source code.

There are tools that facilitate code analysis. As mentioned earlier, each APK file is a ZIP file, and the files in it are stored in a binary form. Therefore, the file content cannot be analyzed directly. The aforementioned Apktool converts `AndroidManifest.xml` into text. It can also generate *smali code*[8] by reverse engineering the bytecode in `classes.dex`. Although the smali code is not the original Java code, it accurately represents the bytecode in a human-readable manner, which is useful for analyzing the bytecode. In fact, an APK decompiled into smali code can be modified and then recompiled into a working APK. In addition, dex2jar[9] can convert a `classes.dex` file into a JAR file, although the JAR file cannot be recompiled into a working APK. There are several other tools available online that are useful for APK file analysis.

A set of APIs are invoked during the runtime of each application. Each API is associated with a particular permission. When an API call is made, the approval of its associated permission is checked. The execution of the API is successful only if the necessary permission is granted by the user. This way, the permissions are used

---

[7] https://source.android.com/devices/tech/dalvik/dex-format

[8] https://github.com/JesusFreke/smali

[9] http://code.google.com/p/dex2jar/downloads/list

to protect the user's private information from unauthorized access [9]. API calls of the Android application are stored in a smali file, which can be obtained through reverse engineering.

7.3 Techniques for Identifying Android Malware

Malware is distributed to Android terminals as APKs. To identify malware, APK analysis techniques are needed. Automated techniques are highly desired for malware analysis. In this section, three types of analysis (blacklisting, parameterizing, and classification) are introduced.

7.3.1 Blacklisting

Blacklist-based detection techniques are often used in a variety of fields, such as spam filtering and malware detection. There are specific blacklists for each application, e.g., blacklists of APK files, blacklists of URLs, and blacklists of application developer signatures. An APK file blacklist is a list of hash values of APK files identified as malware. A URL blacklist enumerates URLs that host malicious contents, such as malware, and APK files that communicate with these URLs identified as malware. A blacklist of application developers is a list of the certificates of malware developers, and it is very likely that APK files with these certificates are indeed malware.

Blacklisting relies on blacklists that have already been created and are readily available. For this reason, alternate means are necessary to evaluate applications that have not been previously evaluated. While manual evaluation is one option, there are also automated approaches, as discussed in the following sections.

7.3.2 Parameterizing

One of the automated approaches to judge whether an application is malware is to define a numerical parameter that represents the likelihood of the software being malware. If the value of this parameter exceeds a certain value, the software is considered malware.

DroidRisk is such a technique [10]. It quantifies the risk level of an application from its permission requests. First, it quantifies the risk level of each permission by multiplying the probability of the permission being misused by malware by the effect of such misuse. Then it sums the quantified risks of the permissions requested by the application as follows to produce a parameter that represents the risk level of the application:

$$r = \sum_i \left\{ L(p_i) \times I(p_i) \right\}, \tag{7.1}$$

where r denotes the quantified risk level, $L(p)$ denotes the likelihood of permission p being used by malware, and $I(p)$ denotes the effect of permission p being misused by the malware. If the value of r exceeds a predefined threshold, DroidRisk labels the application as malware.

This scheme may be able to recognize many malware applications, but its capabilities are limited. For example, application types are not considered. The probability of a permission misuse varies depending on the application type, and this variance is not considered in DroidRisk. For example, there is nothing unusual in a calendar application requesting permission to access the user's contact list, but it is suspicious if a calculator application does so.

There is a scheme that takes into account the application type. It identifies a particular permission, called *category-based rare critical permission (CRCP)*, for each application type [11]. Although this permission could be used by many applications, it is rarely used by applications of a particular type. If the permission is requested by an application of a particular type, this scheme considers the application to be malware.

The schemes in this category are fairly easy for analysts to understand and verify. However, these schemes utilize only a few features. Using of more features can improve the performance of malware detection, but constructing a parameter using multiple features is a non-trivial task. The state-of-the-art classification approach described in the next section can take into account multiple features and can provide better malware identification performance than previous approaches.

7.3.3 Classification

In our classification approach, each sample is classified instead of defining and using key parameters. If malware detection is the only concern, schemes using this approach classify the samples into two groups. One might argue that the schemes described above can also be considered classification schemes, but unlike them, the schemes here usually do not provide human-friendly parameters and use machine learning techniques, which are well-known to be able to outperform other types of techniques in classification tasks.

There are several different machine learning techniques, but not all of them are suitable for classifying Android applications. As described in Sect. 7.4, various features that include API calls are used as SVM input in this chapter. Encoding these features as numerical attributes results in a very large dataset. This is particularly true for the API feature: more than 30,000 unique APIs are used by the APK files in our data. This high-dimensional and sparse data makes it difficult to use common machine learning techniques. However, according to Vapnik's *statistical learning theory* [6], an SVM has guaranteed performance even for extremely high-dimensional data. A linear SVM is particularly preferable for such data because of its fast convergence rate and favorable generalization performance. Another reason to prefer a linear SVM over other methods is, as will be discussed in Sect. 7.7, that it facilitates fast feature selection for high-dimensional data. Therefore, only SVM is used for this purpose in this chapter.

7.4 Dataset Preparation

Before running machine learning, the dataset must be prepared. This section describes the collection of common data types.

7.4.1 APK File Analysis

APK files are required to generate various datasets for analysis. They are available in online APK markets, such as Google Play.[10] There are several options to download these files, and there are tools and APIs specially designed for this purpose that are worth considering [12]. Some markets set several restrictions on the use of and access to APK files, which are described in the corresponding terms and conditions.

By analyzing APK files, the data required to build a dataset can be extracted. There are basically two types of analyses for extracting features: static analysis and dynamic analysis.

Static analysis inspects the files inside an APK file. Among these files, `AndroidManifest.xml` and `classes.dex` contain data suitable to be used as features. Permission request information can be extracted from `AndroidManifest.xml`, and API calls from `classes.dex`. Further information, such as information about the intended application, may also be extracted from `AndroidManifest.xml`, although this chapter utilizes permission request information and API call information only.

In contrast to static analysis, dynamic analysis monitors the behavior and activities of running applications. There are tools that facilitates such analyses, e.g., TaintDroid [13] and Epicc [14]. A common drawback of dynamic analysis is the limited interaction with the user interface, although some tools, such as Monkey [15] and DroidBot [16], address this issue.

While both types of analysis are important, this chapter focuses on static analyses.

7.4.2 Application Metadata

APK analysis is not the only way to generate data for datasets. Other data sources include, for example, the metadata of APK files available on online APK markets. The APK files on APK markets are published with an application description. Application

[10] https://play.google.com

category information and the number of downloads are also often available. These pieces of information can be collected and used as features.

Because application descriptions cannot be handled in their original form, they need to be converted so that they can be used as SVM input. One way to do this is to apply the bag-of-words model, which lists the frequencies of each word appearing in the description [17]. However, the bag-of-words model generates high-dimensional datasets, so for this purpose it is inefficient.

Another approach is to generate application clusters from the descriptions by applying a clustering algorithm, such as k-means [18], to the generated bag of words. CHABADA, for example, uses k-means to define clusters and then classifies the APK files into clusters [19]. This process consists of three stages:

1. Data preprocessing stage. Words usable for latent Dirichlet allocation (LDA) [20] are produced from the descriptions. First, the description language is checked and the non-English descriptions are discarded. Non-textual items, i.e., numbers, HTML tags, web links, and email addresses, are also discarded. Then, stems[11] are extracted from the descriptions and stop words are truncated. Finally, the number of words in the final description is counted. All descriptions that contain less than ten words are discarded.[12]
2. Topic model generation stage. The words in the remaining descriptions are processed using LDA. These descriptions are imported and a number of topics are trained. A total of 300 topics were considered, with a topic proportion threshold of 0.05 and a maximum of four topics per entry. As a result, this process outputs several (maximum four) topic number-proportion value pairs.[13]
3. Cluster generation stage. The APK files are grouped into clusters in accordance with the topic number-proportion value pairs for each description using k-means.[14] The number of categories was set to 12, which is identical to the number of categories used in the Opera Mobile Store.[15]

7.4.3 Label Assignment

To run supervised learning, label information is needed first. There are several techniques for obtaining this information, one of which is introduced here. It uses VirusTotal,[16] an information aggregator that derives data from the output of various eval-

[11] A stem is a part of a word and is common to all its inflected forms.

[12] We used the language detection library [21] to detect the language, stemmify [22] for the stemming operation, and the stoplist of MALLET [23] as the list of stop words.

[13] We used MALLET for running LDA and considered 300 topics because the MALLET documentation states that "the number of topics should depend to some degree on the size of the collection, but 200–400 will produce reasonably fine-grained results."

[14] We used the "kmeans" function of Ruby gem [24].

[15] Opera Mobile Store was rebranded and is now called Bemobi Mobile Store [25].

[16] https://www.virustotal.com

uation engines. If at least two of the results indicate that the file is malicious, the APK file is considered malware.

One issue to address is precisely defining malware. Adware is a software type that is merely a nuisance (does not pose a threat) and has characteristics different from those of malware. While adware can be considered malware as well, for our purposes, adware needs to be distinguished from malware. Therefore, any malware applications with a name that includes the string "adware" or any of the strings listed in an adware family name list should be removed. The adware family name list can be built in different ways, one of which is listing all the software names identified as adware by VirusTotal. Note that the naming rules may differ depending on the evaluation engine, which needs to be considered when building the list.

7.4.4 Data Encoding

Encoding information from various sources into numerical features can be challenging: feature format and availability of features may vary from source to source. For example, while the weight and order may carry essential information for API calls, they are unavailable for features like permission requests and application categories.

To encode all features consistently, we encode all the features as binary attributes. For permission requests and API calls, an attribute is set to 1 if the permission/API is declared in the manifest/smali file; otherwise it is set to 0. Application categories can be encoded as binary attributes following *one-hot encoding*: each application category is modeled as a binary attribute, and the attributes are mutually exclusive so that only one in the set can have a value of 1. To encode description text into binary attributes, we first perform clustering analysis using the technique mentioned in Sect. 7.4.2 so that each of the APKs is assigned to one of a handful of clusters. After that, encoding can be done following one-hot encoding. Hence, essential discriminant information—the more similar the description text of two APK files, the more likely they fall into the same class—is encoded as binary attributes.

Exploitation of description text using advanced text mining methods might lead to better generalization performance. Nevertheless, it also introduces new problems, such as significantly increased data dimensionality and the reliance on feature weighting, so this approach is omitted here.

7.4.5 A Novel Dataset of Safe and Malicious APK Files

We generated our dataset[17] using the techniques described in the previous section. We collected 87,182 APK files from the Opera Mobile Store for the period January–September 2014. The files from which permission requests could not be extracted were excluded simply because permission requests are strictly required for our ana-

[17]http://mobilesec.nict.go.jp

lyzing schemes. The files that VirusTotal could not handle were also excluded from the dataset, because VirusTotal evaluation results are needed to label the dataset. Following the procedure described in Sect. 7.4.3, the files were labeled as malicious or safe, and adware was omitted. The result was a dataset of 61,730 APK files, consisting of 49,045 safe and 12,685 malicious files.

We also collected APK file metadata from the Opera Mobile Store for the same period. This metadata included the application category, the description, and the number of downloads. The description was used to generate the cluster information. The breakdowns of the dataset by category and cluster are shown in Tables 7.2 and 7.3, respectively. All the collected data were encoded using the procedure described in Sect. 7.4.4.

Table 7.2 Dataset by category

Category	Safe	Malicious	Total
Business and finance	3,779	268	4,047
Communication	2,114	323	2,437
E-books	2,784	479	3,263
Entertainment	14,138	2,453	16,591
Games	12,090	2,603	14,693
Health	1,536	228	1,764
Languages and translators	734	41	775
Multimedia	2,422	567	2,989
Organizers	1,300	87	1,387
Ringtone	327	132	459
Theme skins	5,276	5,059	10,335
Travel and maps	2,545	445	2,990
Total	49,045	12,685	61,730

Table 7.3 Dataset by cluster

Cluster	Safe	Malicious	Total
1	3,574	934	4,508
2	3,883	889	4,772
3	3,945	976	4,921
4	5,247	1,206	6,453
5	4,317	1,174	5,491
6	3,820	1,077	4,897
7	3,474	919	4,393
8	5,337	2,091	7,428
9	4,104	811	4,915
10	4,346	832	5,178
11	3,496	818	4,314
12	3,502	958	4,460
Total	49,045	12,685	61,730

7.5 Detecting Malware Using SVM

In this section, an SVM is used with the generated dataset to detect malware. Different feature schemes are used to achieve better performance. The SVM as well as well-established techniques and metrics for performance evaluation are also explained.

7.5.1 SVM: A Brief Overview

An SVM, as described in previous chapters, is a machine learning model that maps features of data samples and draws a decision hyperplane that divides them into two groups (see Fig. 7.2).

More precisely, an SVM draws hyperplanes (the dotted lines in the figure) that crosses the outermost sample for each of the two groups, and draws another hyperplane (the solid line) that maximizes the distance to each of the hyperplanes. The hyperplane serves as the borderline between the two groups. When applied to Android malware identification, an SVM can divide APK files into two groups by using the features of the APKs and drawing a decision hyperplane between the features of safe software (benignware) and those of malware.

The idea of a two-class SVM is described as follows. From a set of training samples $\mathcal{D} = \{(\boldsymbol{x}_i, y_i) | \boldsymbol{x}_i \in \mathbb{R}^d, y_i \in \{-1, +1\}, i = 1, \ldots, \ell\}$, a two-class SVM learns a norm 1 linear function,

$$f(\boldsymbol{x}) = \langle \boldsymbol{w}, \boldsymbol{x} \rangle + b, \tag{7.2}$$

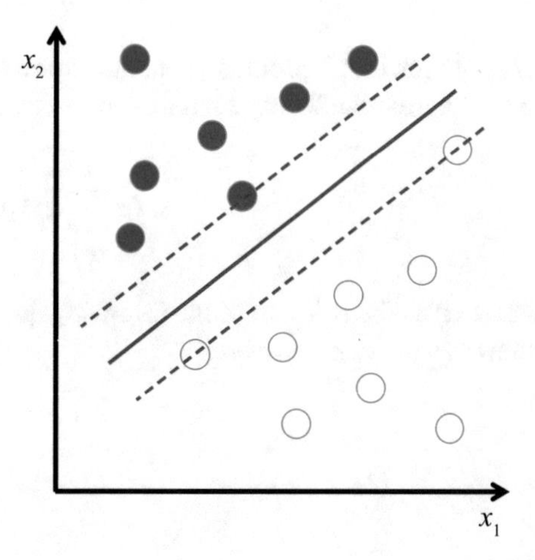

Fig. 7.2 Division of samples with 2-dimensional features into two groups using an SVM

determined by weight vector $\boldsymbol{w}$ and threshold b, which represents the maximum margin.[18] According to Vapnik's statistical learning theory, margin maximization in the training set is equivalent to the minimization of the generalization error of the classifier.

If the samples cannot be separated by using a linear hyperplane, there are two options. The first option is to use a penalty parameter, C, and a group of slack parameters, $\xi_i (\geqslant 0)$, to control the trade-off between the magnitude of the margin and the error introduced by the non-separable samples. The second option is to first map the input vector $\boldsymbol{x}_i$ into a high-dimensional (possibly infinite) feature space, $\mathcal{F}$, through a nonlinear mapping function Φ, such that the improved separability between the training samples from opposite classes could ensure the existence of feasible solutions of the maximum margin hyperplane in $\mathcal{F}$. With the so-called *kernel trick*, mapping Φ could be implicitly implemented by some kernel function $K(\cdot, \cdot)$, which corresponds to an inner product in the feature space, i.e.,

$$K(\boldsymbol{x}_i, \boldsymbol{x}_j) = \langle \Phi(\boldsymbol{x}_i), \Phi(\boldsymbol{x}_j) \rangle. \tag{7.3}$$

In practice, it is common to use both options in the learning phase to gain numerical robustness (from the first option) and adaptivity to nonlinear problems and improved separability (from the second option) [26].

For a kernel-based SVM with misclassified samples being linearly penalized with a weight parameter C, usually referred to as the *soft margin parameter*, the optimization problem can be written as

$$\begin{cases} \min \ \dfrac{1}{2}\|\boldsymbol{w}\|^2 + C\sum\limits_{i=1}^{\ell} \xi_i, \\ s.t. \ \ y_i f(\Phi(\boldsymbol{x}_i)) \geqslant 1 - \xi_i, \\ \qquad \xi_i \geqslant 0, \quad i = 1, \ldots, \ell. \end{cases} \tag{7.4}$$

The solution to this problem, generally known as the *primal problem*, can be obtained using Lagrangian theory such that $\boldsymbol{w}$ can be computed as

$$\boldsymbol{w} = \sum_{i=1}^{\ell} \alpha_i^* y_i \Phi(\boldsymbol{x}_i), \tag{7.5}$$

where α_i^* is the solution of the following quadratic optimization problem, generally known as the *dual problem*:

[18] A margin is the distance from the hyperplane to the nearest training data point of either class.

$$\begin{cases} \max\ W(\boldsymbol{\alpha}) = \sum_{i=1}^{\ell} \alpha_i - \frac{1}{2} \sum_{i,j=1}^{\ell} \alpha_i \alpha_j y_i y_j K(\boldsymbol{x}_i, \boldsymbol{x}_j), \\ s.t.\ \sum_{i=1}^{\ell} y_i \alpha_i = 0, \\ \quad 0 \leqslant \alpha_i \leqslant C, \quad i = 1, \ldots, \ell. \end{cases} \tag{7.6}$$

After training, the SVM predicts the class label of an incoming test sample $\boldsymbol{x}$ on the basis of the following decision function:

$$f(\boldsymbol{x}) = \sum_{i=1}^{\ell} \alpha_i^* y_i K(\boldsymbol{x}, \boldsymbol{x}_i) + b, \tag{7.7}$$

which is a kernelized version of (7.2). If $f(\boldsymbol{x}) > 0$, $\boldsymbol{x}$ is assigned to the positive class, otherwise it is assigned to the negative class.

In our proposed scheme, the linear SVM solver proposed in [27] is used to build the model from the training data. In short, the following L2-SVM is solved using the trust-region Newton method introduced in [27] and implemented in the LIBLINEAR toolbox [28]:

$$\min_{\boldsymbol{w}} g(\boldsymbol{w}) = \frac{1}{2} \boldsymbol{w}^T \boldsymbol{w} + C \sum_{i=1}^{\ell} \Big(\max(0, 1 - y_i \boldsymbol{w}^T \boldsymbol{x}_i) \Big)^2 \tag{7.8}$$

7.5.2 Feature Settings

Several academic papers have reported malware detection schemes using SVM. Peiravian and Zhu [29] extracted permission requests and API calls as features and used machine-learning techniques including an SVM. DroidAPIMiner [30] and Li et al. [31] perform feature selection in order to improve the performance of machine learning techniques.

To classify samples with high accuracy, one needs to choose features that capture the characteristics of benignware and malware. Different types of features could be used, including requested permissions, application categories, application descriptions, and the number of downloads. This chapter explores the use of an SVM for detecting malware by preparing various features and evaluates which types of features contribute effectively to the detection performance.

To investigate the relevance of different features in distinguishing malware from benignware, other evaluation metrics and classification accuracy are compared using different feature settings. Starting with the major features, side features are added one type at a time to see how much performance improvement can be obtained by incorporating new features into the learning.

To identify the various feature schemes, the following notation is used in the next sections. SVM(p) denotes an SVM trained exclusively using the permission request feature p. SVM(p, ct) denotes an SVM trained using p and the application category feature ct. SVM(p, cl) denotes an SVM trained using p and the application cluster feature cl. SVM(p, ct, cl) denotes an SVM trained using p, ct, and cl. Other feature schemes can be formulated the same way; e.g., SVM(p, api, ct, cl) denotes an SVM trained using all three features plus the API feature api.

7.5.3 Hyperparameter Tuning

Algorithm performance often depends heavily on the hyperparameters used to train the SVM model. *Hyperparameter optimization* finds a tuple of hyperparameters that yields to a model that maximizes the generalization performance on independent data. Before performance evaluation, cross-validation is performed to estimate the generalization performance and identify the optimal hyperparameters for later use.

To perform cross-validation, the training set is first randomly partitioned into K disjoint folds (we set $K = 10$ as usual). Then, in the ith iteration, an SVM is trained using the complementary set of the ith fold, and tested against the ith fold. Finally, the evaluation is done by averaging the results over K iterations. For a linear SVM, the only critical parameter is the penalty parameter C in (7.4), which controls the trade-off between the optimization objective and the number of training errors. We perform a *grid search*—simply an exhaustive search through a manually specified subset of C values—to determine the C that produces the best cross-validation result on the training dataset. Parameter C is then used to train an SVM model using all samples in the training set. Finally, the performance is evaluated by comparing the model output against the true labels of the test data.

7.5.4 Evaluation Metrics

There are common metrics for assessing classifier performance. They are calculated using four intermediate parameters: true positive (TP), false positive (FP), true negative (TN), and false negative (FN). TP is the number of predicted positive records classified correctly, FP is the number of predicted positive records classified incorrectly, TN is the number of predicted negative records classified correctly, and FN is the number of predicted negative records classified incorrectly. Based on the four intermediate parameters, four metrics are used for performance assessment:

- *Accuracy*: the probability of test records being classified correctly, i.e.,

$$Accuracy = \frac{TP + TN}{n} \tag{7.9}$$

- *Precision*: the probability of predicted positives being classified correctly, i.e.,

$$Precision = \frac{TP}{TP + FP} \tag{7.10}$$

- *Recall*: the probability that a record is classified correctly, i.e.,

$$Recall = \frac{TP}{TP + FN} \tag{7.11}$$

- *False positive rate (FPR)*: the probability that a negative result is classified incorrectly, i.e.,

$$FPR = \frac{FP}{FP + TN} \tag{7.12}$$

When measuring the performance in experiments, the hyperparameters that result in maximum accuracy are considered optimal. For practical applications that provide security alerts, a scheme that minimizes the false negative rate (FNR) is desired, which is equivalent to $(1 - recall)$, while for applications that provide automated countermeasures, minimizing FPR might be a better choice. Which metric to use has to be determined on a case-by-case basis.

7.5.5 Numerical Results

To evaluate the generalization performance of the classifiers, the dataset can be randomly divided into training and test sets, and the experiment can be conducted several times. In our experiment, 70% of the data were used for training, and the remainder were used for performance evaluation. The reported results average the results of ten rounds.

In Table 7.4, we compare the overall generalization performance using different feature schemes.

Table 7.4 Performance of SVM-based schemes

Feature	Accuracy [%]	Precision [%]	Recall [%]	FPR [%]	Train [s]	Test [μs]
SVM(p)	88.87 ± 0.12	81.19 ± 0.65	59.67 ± 0.41	3.58 ± 0.16	0.08	0.3
SVM(p, ct)	89.45 ± 0.15	83.87 ± 0.60	60.27 ± 0.80	3.00 ± 0.15	0.19	0.3
SVM(p, cl)	89.38 ± 0.10	83.48 ± 0.67	60.25 ± 0.38	3.08 ± 0.16	0.14	0.3
SVM(p, ct, cl)	89.45 ± 0.09	83.76 ± 0.53	60.35 ± 0.59	3.03 ± 0.14	0.31	0.3
SVM(api)	94.07 ± 0.16	87.23 ± 0.45	83.34 ± 1.00	3.16 ± 0.15	45.08	12.4
SVM(api, ct)	94.09 ± 0.17	87.37 ± 0.49	83.26 ± 0.73	3.11 ± 0.14	35.65	12.4
SVM(api, cl)	94.08 ± 0.17	87.23 ± 0.53	83.40 ± 0.79	3.16 ± 0.16	41.33	12.4
SVM(api, ct, cl)	94.07 ± 0.15	87.20 ± 0.38	83.36 ± 0.89	3.16 ± 0.12	40.24	12.4
SVM(p, api)	94.07 ± 0.19	87.19 ± 0.57	83.39 ± 0.80	3.17 ± 0.16	37.94	12.4
SVM(p, api, ct, cl)	94.05 ± 0.18	87.21 ± 0.31	83.28 ± 0.99	3.16 ± 0.10	43.55	12.4

Because accuracy, precision, and recall show similar tendencies, accuracy is used in the following discussion. The upper block of the table shows the results for when the model was initially trained using the permission request feature, and then the results for when the application category feature and the cluster feature were used as well. On our dataset, the permission request feature resulted in an accuracy of 88.87%. The accuracy increased to 89.45% when the application category feature was added and to 89.38% when the cluster feature was added. A comparison of these values indicates that the application category feature may carry slightly more discriminative information than the cluster feature.

The accuracy remained at 89.45% when these two metadata features were used in addition to the permission request feature. In other words, the two metadata features carry nearly equally valuable information when used for malware detection.

The middle block of Table 7.4 shows the results using the API feature. Here, the learning started with only the API feature; the other two metadata features were added later. Compared to the results for SVM(p), SVM(api) shows significant improvement for all four performance criteria. The 94.07% accuracy is a 5% performance gain compared to the accuracy with SVM(p). However, when the metadata features were added, there was only a slight improvement in accuracy. This suggests that metadata features carry little extra discriminative information that could be useful for classification.

The bottom block of Table 7.4 shows the results for SVM(p, api) and SVM (p, api, ct, cl). For both of these feature schemes, accuracy remained at the same level that was obtained with the API feature alone. This confirms that API calls carry more fine-grained discriminative information than permission requests. Because of the associative relationship, permission means no additional discriminative information in terms of performance gain compared to classifiers based on the API feature alone. Compared to SVM(p, api), the accuracy of SVM(p, api, ct, cl) dropped slightly when the metadata features were also used for learning. To sum it up, using metadata features with the API feature provides little additional discriminative information to be used for classification.

The last two columns of Table 7.4 show the training and testing times for all settings. Due to the high performance of LIBLINEAR on mid-scale datasets, computation time cannot be considered a bottleneck neither for training nor for testing. More advanced features can also be included in the learning as long as the generalization performance is improved.

7.6 Comparison with Parameterizing

As mentioned earlier, machine-learning-based approaches outperform other schemes in classification. To understand the performance advantage, this section provides a comparison between a parameterizing approach and a machine learning approach (using DroidRisk and SVM, respectively). To enable a fair comparison, DroidRisk was extended to be able to incorporate features beyond permission requests. The following sections first describe the extensions and then compare the performance to that of the SVM-based scheme.

7.6.1 *Extending DroidRisk*

We define a novel scheme, referred to as DR(p, ct), which extends DroidRisk to quantify security risks of APK files based on their application category ct. DroidRisk determines $L(p)$ and $I(p)$ by analyzing the entire dataset, but the optimal values can differ depending on the application type. For example, many applications in the "Travel&maps" category request permission for using GPS, whereas few within the "Multimedia" category request the same. Thus, DR(p, ct) sets different values for $L(p)$ and $I(p)$ for each application category by analyzing the data in each category as follows:

$$r = \sum_i \left\{ L(p_i, ct) \times I(p_i, ct) \right\}. \quad (7.13)$$

We also define another scheme, referred to as DR(p, cl), which extends DroidRisk, and uses application cluster cl, instead of ct, where cl is derived using the scheme described earlier in Sect. 7.4.2. Similar to the category-based scheme, this scheme calculates the risk value r as

$$r = \sum_i \left\{ L(p_i, cl) \times I(p_i, cl) \right\}. \quad (7.14)$$

7.6.2 *DroidRisk Performance*

Table 7.5 shows the performance of DroidRisk and its extensions defined in the previous section in terms of accuracy, precision, recall, and FPR.

Table 7.5 Performance of DroidRisk (DR)-based schemes

	Accuracy [%]	Precision [%]	Recall [%]	FPR [%]
DR(p)	83.59±0.14	67.02±0.61	39.65±1.29	5.05±0.25
DR(p, ct)	85.63±0.20	59.68±2.19	29.85±1.69	4.93±0.48
DR(p, cl)	83.88±0.17	65.78±0.88	42.35±1.75	6.05±0.40
DR(api)	79.50±0.04	51.77±18.47	0.93±0.34	0.18±0.06
DR(api, ct)	82.41±0.12	45.82±9.01	14.07±0.70	6.75±0.86
DR(api, cl)	79.53±0.04	53.84±9.17	0.85±0.26	0.13±0.04

The values are the averages of 10×10 cross-validation results. As shown in the table, the performance of DR(p) was improved by the use of application categories and clusters. Compared to application clusters, application categories improve the performance further, although the contribution of application clusters could be improved even further by fine-tuning the cluster generation algorithms. However, the performance improvement is insignificant. Most probably this is because both DR(p, ct) and DR(p, cl) suffer from the overfitting caused by dividing the dataset for contextual-based analysis.

One might argue that API calls should be used instead of permission requests for more accurate analysis. To investigate this, we measured the performance of these schemes using API calls instead of permission requests. More than 30,000 types of API calls, including Android Framework APIs, Java APIs, and third party APIs, were analyzed to calculate parameters L and I. For the sake of simplicity, the value of L was optimized for the top 10% of API calls used by malware, while the value was set to 1 for the remainder. Table 7.5 shows the performances of DR(api), DR(api, ct), and DR(api, cl), where api denotes API calls. The performances of these API-based schemes lag well behind those of the permission-based schemes, most probably because the API-based schemes are affected more by the overfitting problem than the permission-based ones. The degree of overfitting is determined by the number of samples in the dataset and the dimensionality of features. Because the dataset here is the same, overfitting has a higher impact on those API-based schemes that use a higher dimensionality of features than on the permission-based schemes. When using API-calls and metadata, the degree of overfitting has to be taken into account.

A comparison of the results in Tables 7.4 and 7.5 clearly shows that SVM-based schemes provide better performance than DroidRisk-based ones. All of the DroidRisk-based schemes provide better performance than a random classifier. In fact, they could be regarded as weak classifiers. These results confirm the observation made earlier in Sect. 7.3 that classification approaches outperform parameterizing approaches.

7.7 Feature Selection

The four feature types described in Sect. 7.4, when encoded as binary attributes, yield 37,720-dimensional feature vectors (583 dimensions for permission feature, 37,113 for API feature, 12 for category feature, and 12 for cluster feature). However, according to the evaluation results presented in Sect. 7.5.5, not all binary attributes contribute discriminative information equally for classification. This observation gives rise to a more rigid analysis of the relevance of features. In the following, except when used in the context of *feature selection*, we use *attribute* to denote a single element of the vector input to the classifier. We use *feature* to refer to a group of variables obtained from a particular source.

In machine learning, feature selection is a well-explored tool to 1) defy the curse of dimensionality for generalization performance gain, 2) speed up computation in training and prediction, 3) reduce data gathering and storage cost, and 4) facilitate further data investigation. Traditional feature selection methods usually fall into one of two categories: wrapper methods or filter methods. Recently, more advanced methods have also been introduced, for example embedded methods, which add selection criteria to the objective functions, and hybrid methods, which combine multiple selection strategies.

Previous works on feature selection suggest that feature selection is essential to build strong and robust classifiers. This also holds for SVMs, which specialize in handling high-dimensional data. However, feature selection is often computationally intensive: it usually requires repeated training of the same classifiers with different feature subsets. Considering the large number of APK files and the high dimensionality of our data, a cost-effective feature selection method is needed.

7.7.1 Recursive Feature Elimination

Thanks to the soundness and high efficiency of SVMs, several SVM-based feature selection algorithms have been proposed. Among them, recursive feature elimination based on SVM (SVM-RFE) [32] has become widely used for analyzing large-scale, high-dimensional data.

SVM-RFE was originally introduced to select relevant genes for cancer classification. It uses the backward selection logic: it starts with fitting an SVM model with all the attributes of interest, then ranks the attributes in accordance with the criterion described in the next section, with those that are insignificant at a chosen critical level omitted. Next, the algorithm successively refits reduced SVM models using the remaining attributes and applies the same rule until all remaining attributes are statistically significant or until a predefined number of attributes remain. A modified version of RFE [32] is shown in Algorithm 11.

Algorithm 11: SVM recursive feature elimination

Data: $\mathcal{D} = \{(\boldsymbol{x}_i, y_i)\}$ of dimension D
Result: $\mathcal{R}$ (Ranked attribute indices with increasing significance)
1 $\mathcal{A} = \{1,...,D\}$ (Default attribute index set);
2 **while** $|\mathcal{A}| \geqslant 1$ **do**
3 | $SVM(\alpha_i^*)$ = Train($\mathcal{A}, \mathcal{D}$);
4 | $R_k = | \sum_{i=1}^{\ell} y_i \alpha_i^* x_{ik} |$, for $k \in \mathcal{A}$;
5 | $e = \arg\min_k R_k$;
6 | Push($\mathcal{R}, e$);
7 | $\mathcal{A} := \mathcal{A} - e$;
8 **end**

7.7.2 Ranking Criterion

A cost-effective ranking criterion was introduced in [32] for estimating the contribution of an attribute to the classification. Given the optimal solution of the objective function for the linear SVM in (7.8) as $\boldsymbol{\alpha}^*$ and $\boldsymbol{w}^*$, we have

$$\boldsymbol{w}^* = \sum_{i=1}^{\ell} y_i \alpha_i^* \boldsymbol{x}_i \tag{7.15}$$

Equation (7.15) shows that if some elements of $\boldsymbol{w}^*$ are zero, deletion of the associated input attributes will keep the decision function unchanged. Moreover, an attribute associated with an element close to zero in $\boldsymbol{w}^*$ may be considered insignificant—its removal could be done without degrading the generalization ability of the classifier. Therefore, the ranking criterion of the kth feature for a linear SVM can be defined as

$$R_k = \sqrt{\|\boldsymbol{w}^*\|^2 - \|\boldsymbol{w}^{*(k)}\|^2} = |\sum_{i=1}^{\ell} y_i \alpha_i^* x_{ik}| \tag{7.16}$$

where x_{ik} is the kth element of $\boldsymbol{x}_i$, and $\boldsymbol{w}^{*(k)}$ is obtained from $\boldsymbol{w}^*$ by setting all components x_{ik} to 0 for $i = 1, \ldots, \ell$.

7.7.3 Experiment

We ran SVM-RFE on our dataset using all four feature types described in Sect. 7.4 to find attributes that produce optimal performance. Since there are trade-offs between different types of performance metrics, we chose the F-measure as the metric to be prioritized. It is the harmonic mean of precision and recall:

$$F\text{-}measure = 2 \cdot \frac{precision \cdot recall}{precision + recall}. \tag{7.17}$$

Note that an other metric could be prioritized, as discussed in Sect. 7.5.4.

To speed up the feature selection, when SVM-RFE is applied to our data, we eliminate multiple attributes at each iteration. Let the total number of attributes before feature selection to be D. Let the dimension of the binary input vector of SVM and the number of attributes removed at the ith iteration to be d_i and Δ_i, respectively. In the experiment, Δ_i is determined as follows:

Table 7.6 Performance of SVM-RFE

No.	D_{opt}	F-Measure	Accuracy [%]	Precision [%]	Recall [%]	FPR [%]	D_{base}
1	3,129	0.857	94.18	86.62	84.76	3.38	878
2	17,115	0.868	94.59	88.15	85.50	3.02	1,439
3	3,885	0.853	94.02	87.40	83.23	3.15	2,734
4	34,112	0.854	94.08	86.82	84.03	3.31	4,554
5	1,927	0.854	94.04	87.15	83.74	3.25	666
6	2,542	0.853	94.19	87.75	83.01	2.95	1,471
7	25,869	0.857	94.21	86.54	84.85	3.39	3,397
8	18,000	0.852	94.03	86.97	83.50	3.24	484
9	10,329	0.853	94.11	86.78	83.81	3.26	7,776
10	1,734	0.855	94.07	87.22	83.83	3.23	808
11	8,918	0.854	94.14	86.93	84.01	3.25	963
Average	11,596	0.855	94.15	87.12	84.02	3.22	2,288
Median	8,918	0.854	94.11	86.97	83.83	3.25	1,439

$$\Delta_i = \begin{cases} \lfloor 0.10 d_i \rfloor, & d_i > 0.80D, \\ \lfloor 0.005 d_i \rfloor, & 100 < d_i \leqslant 0.80D, \\ 1, & d_i \leqslant 100. \end{cases} \tag{7.18}$$

In this experiment, SVM-RFE was run 11 times on our dataset. Table 7.6 shows the performances for each run. The column "D_{opt}" shows the number of attributes needed to produce the optimal performance. The F-measure, accuracy, precision, recall, and FPR of an SVM using the optimal feature set are also shown. A comparison between Tables 7.4 and 7.6 clearly shows that SVM performance is further improved when selecting relevant attributes and omitting irrelevant attributes with SVM-RFE.

The column "D_{base}" in Table 7.6 shows the number of attributes needed to achieve the same performance achieved by an SVM without feature selection, which is shown in Table 7.4. As can be seen, only a median of 1,439 attributes (7,776 at maximum) are needed to achieve the same performance.

To further analyze the results of the experiment, the second run of SVM-RFE on our dataset providing the median value of D_{base} was analyzed further. Figure 7.3 shows the result of the second run of SVM-RFE on our dataset.[19] As can be seen in the figure, the performance drastically changes when we add contributing attributes in the early stage, and the curves saturate fast: the performance barely changes when extra attributes are added beyond 2,000 attributes in the training.

By using a limited number of attributes (a median of 8,918 and a maximum of 34,112), the performance is improved further. The same level of performance as for an SVM without feature selection can be achieved by using an even smaller number of attributes (a median of 1,439 and a maximum of 7,776), which would

[19]The shapes of the curves were similar for all the runs.

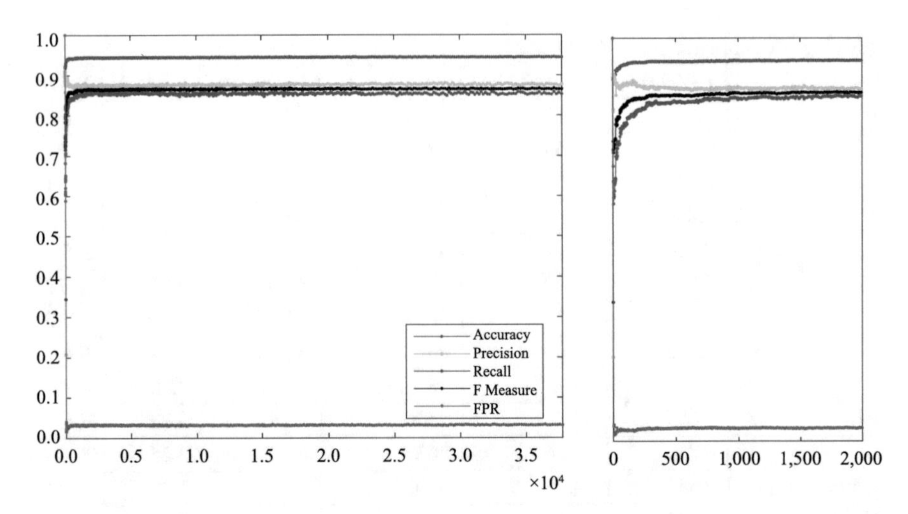

Fig. 7.3 Feature selection results (left: for all attributes, right: for top attributes)

drastically decrease the complexity of calculation. Therefore, ranking attributes and selecting contributing attributes is an effective way to achieve better performance while minimizing computation cost.

A closer look into the ranked attributes in the results for the second run reveals that API calls are the most useful features in this domain. The first appearances of permission requests, API calls, categories, and clusters in the feature ranking were 14th, 1st, 240th, and 942nd, respectively. Because all of these features appeared before the dimension reaches D_{opt} at least once, they contribute to malware detection. However, the earlier they appear, the more contributing they are. These results support the results presented in Sect. 7.5.5—API calls contribute the most while cluster information contributes little to the classification. The results for the second run also revealed that some attributes of permission request feature and application category feature contributes to classification performance.

7.8 Issues and Limitations

Android application analysis is important for securing Android devices, and machine learning can efficiently assist in doing this. However, the following issues and limitations need to be considered:

1. *False positives and false negatives.* Machine learning techniques can perform classification with good accuracy and precision, but false positives and false negatives are inevitable. These can cause a significant risk in real-world applications. To minimize this risk, additional measures are needed. For example, the classification results of machine learning should be reviewed by automated security

tools and, in some cases, by human operators to produce whitelists and blacklists of safe and dangerous applications, which can be offered to users to be used as alternates to classification results.
2. *Dataset preparation.* Android application analysis requires large enough datasets to be able to extract effective statistics and features, and to evaluate the effectiveness of assorted classification algorithms. Online data collection has barriers such as copyright and usage restrictions. The use of these data sources is therefore determined by the corresponding legislations, terms, and conditions.
3. *Criteria for labeling malware.* When running a machine learning algorithm, labeling is essential. In this chapter, applications have been classified as malware or benignware, and this classification was based on labeling. Labeling each and every entry in the dataset as malicious or safe is not always easy because malware characteristics are not trivial to define. For example, it is debatable whether adware is malware or not. As a result, some malware analysis engines in VirusTotal treat adware as malware while other engines do not. Depending on the definition of malware, labeling results may change, which in turn affects the classification results of machine learning. This makes it necessary to have an adequate malware definition for each application.

7.9 Conclusions

In this chapter, techniques for identifying Android malware have been described, and the usability of machine learning techniques for this purpose have been demonstrated, with the emphasis on SVMs. The dataset used for running an SVM to identify malware was generated from permission requests, API calls, application categories, and descriptions. Our evaluation measured the performance of SVM and DroidRisk on our dataset, leading to the conclusion that an SVM has a clear advantage over DroidRisk. To further improve SVM performance, SVM-RFE was used to evaluate attributes and identify their contributions. Removing the non-contributing features resulted in an accuracy of 94.15%. The evaluation confirmed that API calls are the most dominant features, and that permission requests and application categories also make some contributions to classification performance, while application clusters derived from application descriptions barely affect classification. This chapter focused on SVM, but other machine learning techniques, such as deep learning, can also be considered for identifying Android malware. Different types of features and feature selection schemes could also be explored. Such research efforts would greatly contribute to a more secure Android ecosystem.

References

1. Van der Meulen R, Forni AA (2017) Gartner says demand for 4G smartphones in emerging markets spurred growth in second quarter of 2017. https://www.gartner.com/newsroom/id/3788963
2. Sophos (2017) SophosLabs 2018 malware forecast. https://www.sophos.com/en-us/en-us/medialibrary/PDFs/technical-papers/malware-forecast-2018.pdf
3. Lipovsky R (2014) ESET analyzes Simplocker—first Android file-encrypting, TOR-enabled ransomware. https://www.welivesecurity.com/2014/06/04/simplocker/
4. Stefanko L (2015) Aggressive Android ransomware spreading in the USA. https://www.welivesecurity.com/2015/09/10/aggressive-android-ransomware-spreading-in-the-usa/
5. Schölkopf B, Smola AJ (2001) Learning with kernels: support vector machines, regularization, optimization, and beyond. MIT Press, Cambridge
6. Vapnik VN (1998) Statistical learning theory. Wiley, Hoboken
7. Takahashi T, Ban T, Tien CW, Lin CH, Inoue D, Nakao K (2016) The usability of metadata for Android application analysis. In: Hirose A, Ozawa S, Doya K, Ikeda K, Lee M, Liu D (eds) Proceedings of the 23nd International Conference on Neural Information Processing. Springer, Cham, pp 546–554. https://doi.org/10.1007/978-3-319-46687-3_60
8. Ban T, Takahashi T, Guo S, Inoue D, Nakao K (2016) Integration of multimodal features for Android malware detection based on linear SVM. In: Proceedings of the 11th Asia Joint Conference on Information Security, IEEE, pp 141–146. https://doi.org/10.1109/AsiaJCIS.2016.29
9. Moonsamy V, Rong J, Liu S (2014) Mining permission patterns for contrasting clean and malicious Android applications. Future Gener Comp Syst 36:122–132. https://doi.org/10.1016/j.future.2013.09.014
10. Wang Y, Zheng J, Sun C, Mukkamala S (2013) Quantitative security risk assessment of Android permissions and applications. In: Wang L, Shafiq B (eds) Data and applications security and privacy XXVII. Springer, Heidelberg, pp 226–241. https://doi.org/10.1007/978-3-642-39256-6_15
11. Sarma BP, Li N, Gates C, Potharaju R, Nita-Rotaru C, Molloy I (2012) Android permissions: a perspective combining risks and benefits. In: Atluri V, Vaidya J (eds) Proceedings of the 17th ACM Symposium on Access Control Models and Technologies. ACM, New York, pp 13–22. https://doi.org/10.1145/2295136.2295141
12. Demiroz A (2018) Google play crawler JAVA API. https://github.com/Akdeniz/google-play-crawler
13. Enck W, Gilbert P, Han S, Tendulkar V, Chun BG, Cox LP, Jung J, McDaniel P, Sheth AN (2014) TaintDroid: an information-flow tracking system for realtime privacy monitoring on smartphones. ACM T Comput Syst 32(2), Article 5. https://doi.org/10.1145/2619091
14. Octeau D, McDaniel P, Jha S, Bartel A, Bodden E, Klein J, Le Traon Y (2013) Effective inter-component communication mapping in Android with Epicc: an essential step towards holistic security analysis. In: Proceedings of the 22nd USENIX Conference on Security. USENIX Association, Berkeley, CA, USA, pp 543–558. https://www.usenix.org/system/files/conference/usenixsecurity13/sec13-paper_octeau.pdf
15. Android Developers (2018) UI/Application exerciser monkey. https://developer.android.com/studio/test/monkey
16. Li Y, Yang Z, Guo Y, Chen X (2017) DroidBot: a lightweight UI-guided test input generator for Android. In: Uchitel S, Orso A, Robillard M (eds) Proceedings of the 39th International Conference on Software Engineering Companion. IEEE Computer Society, Los Alamitos, CA, USA, pp 23–26. https://doi.org/10.1109/ICSE-C.2017.8
17. Harris ZS (1954) Distributional structure. WORD 10(2–3):146–162. https://doi.org/10.1080/00437956.1954.11659520

18. MacQueen J (1967) Some methods for classification and analysis of multivariate observations. In: Le Cam LM, Neyman J (eds) Proceedings of the Fifth Berkeley Symposium on Mathematical Statistics and Probability, vol 1. University of California Press, Berkeley, pp 281–297. https://projecteuclid.org/euclid.bsmsp/1200512992
19. Gorla A, Tavecchia I, Gross F, Zeller A (2014) Checking app behavior against app descriptions. In: Jalote P, Briand L, van der Hoek A (eds) Proceedings of the 36th International Conference on Software Engineering. ACM, New York, pp 1025–1035. https://doi.org/10.1145/2568225.2568276
20. Blei DM, Ng AY, Jordan MI (2003) Latent Dirichlet allocation. J Mach Learn Res 3:993–1022
21. Shuyo N (2010) Language detection library for Java. https://github.com/shuyo/language-detection
22. Pereda R (2011) Stemmify 0.0.2. https://rubygems.org/gems/stemmify
23. McCallum AK (2002) MALLET: a machine learning for language toolkit. http://mallet.cs.umass.edu
24. RubyGems.org (2013) kmeans 0.1.1. https://rubygems.org/gems/kmeans/
25. Apps and Games AS (2018) Bemobi mobile store. http://apps.bemobi.com
26. Cover TM (1965) Geometrical and statistical properties of systems of linear inequalities with applications in pattern recognition. IEEE Trans Electron EC-14(3):326–334. https://doi.org/10.1109/PGEC.1965.264137
27. Lin CJ, Weng RC, Keerthi SS (2007) Trust region Newton methods for large-scale logistic regression. In: Ghahramani Z (ed) Proceedings of the 24th International Conference on Machine Learning. ACM, New York, pp 561–568. https://doi.org/10.1145/1273496.1273567
28. Fan RE, Chang KW, Hsieh CJ, Wang XR, Lin CJ (2008) LIBLINEAR: a library for large linear classification. J Mach Learn Res 9:1871–1874
29. Peiravian N, Zhu X (2013) Machine learning for Android malware detection using permission and API calls. In: Bourbakis N, Brodsky A (eds) Proceedings of the 25th International Conference on Tools with Artificial Intelligence. IEEE Computer Society, Los Alamitos, CA, USA, pp 300–305. https://doi.org/10.1109/ICTAI.2013.53
30. Aafer Y, Du W, Yin H (2013) DroidAPIMiner: mining API-level features for robust malware detection in Android. In: Zia T, Zomaya A, Varadharajan V, Mao M (eds) Security and privacy in communication networks. Springer, Cham, pp 86–103. https://doi.org/10.1007/978-3-319-04283-1_6
31. Li W, Ge J, Dai G (2015) Detecting malware for Android platform: an SVM-based approach. In: Qiu M, Zhang T, Das S (eds) Proceedings of the 2nd International Conference on Cyber Security and Cloud Computing. IEEE Computer Society, Los Alamitos, CA, USA, pp 464–469. https://doi.org/10.1109/CSCloud.2015.50
32. Guyon I, Weston J, Barnhill S, Vapnik V (2002) Gene selection for cancer classification using support vector machines. Mach Learn 46(1–3):389–422. https://doi.org/10.1023/A:1012487302797

Printed in the United States
By Bookmasters